STUDENT STUDY GUIDE

to accompany

APPLIED CALCULUS

SECOND EDITION

Deborah Hughes-Hallett
University of Arizona

Andrew M. Gleason
Harvard University

Patti Frazer Lock
St. Lawrence University

Daniel E. Flath
University of South Alabama

et al.

Prepared by

Ansie Harding
University of Pretoria
South Africa

JOHN WILEY & SONS, INC.

COVER PHOTO © Lester Lefkowitz/Corbis Stock Market.

To order books or for customer service call 1-800-CALL-WILEY (225-5945).

ISBN 0-471-21361-6

Printed in the United States of America

10 9 8 7 6 5 4 3 2 1

Printed and bound by Victor Graphics, Inc.

To Paul, Katherine and Eleanor - my own family

PREFACE

"Education is not received, it is achieved."

To the student

- The textbook which is supported by this study guide, is not simply a book of formulas and symbols, it is a book of ideas. The emphasis in the textbook is on understanding and applying concepts rather than on using formulas.

- Unfortunately it takes longer to grasp an idea than it does to simply use a formula. To grasp the full meaning of a concept requires of you to read, to think and to question.

- Although a classroom situation serves as an excellent introduction to a topic, it takes more than that to grasp the full implications of a concept. You need to spend some time on making these concepts your own, and to prepare yourself for problem solving.

- Reading a math textbook is a skill that has to be cultivated.

- Use this study guide after a day's lectures and before you start doing problems on the day's Calculus topic. Make sure you know the notation, the concepts and above all what the key points of the day's topic are.

- This study guide is incomplete; it needs YOU to make it complete.

More about the study guide

This study guide is not a short synopsis of the textbook. It is a guide to assist you in systematically working through the textbook, checking whether you truly have mastered the key concepts. It will also help you pick up the important detail that is so easily missed. **Fill-in statements** form the basis of the study guide. One doesn't feel quite as threatened by an incomplete statement as one does by a direct question. This guide also provides ample opportunity to verbalize your understanding of the concepts.

The structure of the study guide

This study guide structures every section as follows:

1. **Key points.** The key points in each section are listed and this will guide you to an overall picture of the study matter.

2. **Knowing the book.** Essential detail from the book is extracted in a guided way. You will sometimes be guided to a particular sentence in the book that contains an essential concept. You can also check whether you know the definitions. DO NOT simply copy from the book.

3. **Checking the concepts.** You can check whether you fully grasp the key points in this section. Having extracted the essential concepts from the book, this is a self-check and a chance to improve your insight.

4. **Checking the examples.** Every example in the textbook is included with a purpose in mind. Selected examples are discussed in review fashion to indicate the significance of the particular example. Working through examples, repeating the examples with different data and answering questions on examples are all invaluable in the process of learning.

5. **Take note** Useful hints and emphasized points are scattered throughout under this heading.

6. **Problems.** Finally, problems are selected for you to solve. Having worked through a particular section you should be ready for and enjoy doing problems.

The ILUO-principle

An idea that has its origin in Japanese industry is that the mastering of a basic skill follows a so-called ILUO pattern. The four letters indicate four different ratings, depending on the level to which a certain skill has been mastered. The lowest rating is an **I** rating and the highest rating is an **O** rating. The reason for the choice of letters is clear from the progression of the four shapes. This rating system can be applied to the process of mastering Calculus:

I

The classroom situation is the student's introduction to a new topic. A few basic concepts are introduced. If the student has followed what was said in class it would earn him an **I** rating. This is the **INITIAL** stage of the learning process. The student has by no means grasped the full implications of what he has learned, nor has he read the textbook. Hopefully he has a basic understanding of the new concepts but he has not been confronted with the finer points and he has not really been able to make this knowledge his own. It is an unfortunate truth that most lectures, however excellent, only serve as an introduction to a particular topic. A good comparison is that of baking a loaf of bread. The initial phase is simply the mixing of ingredients. New ideas are mixed in with the student's existing knowledge.

L

The next leg in the process, the **L** phase, starts when the student reviews what he has learned in class. He should now spend some time on studying the concepts to make it his own by reading the textbook and thinking and talking about the concepts. If he does this as thoroughly as he should then he is bound to come to a better understanding of the concepts and gain insight. He should ideally be confronted with "What if?" type of questions in his mind which could be clarified by re-reading a particular paragraph in the textbook. He should also make notes of the most important concepts and spend some time on examples. Unfortunately this most important phase is also the most neglected phase. Looking at our bread comparison: The second phase is the phase of **LEAVENING**. The bread is left in a warm place and although nothing is added, the bread rises which is essential for the next phase. Similarly the student's knowledge should be kept active and although no new knowledge is added, his insight should increase by studying the concepts. Because most students find it so difficult to read a textbook, they should be guided through this phase. Support material such as this study guide and a good tutor system is advisable. This phase should add to the student's confidence level when confronted with problems in the next phase. It is this phase that is often lacking and where this study guide especially has a role to fulfill.

The third leg in the process should be entered when the student's knowledge is mature enough to answer penetrating questions on the concepts in the section. Most students try to solve problems before they've mastered the concepts properly. They often use the problems as a point of departure and page through the book trying to find something, a formula or an example, to help them solve the problem. In other words, if no problem on a particular concept is given, the student could remain unaware of its existence, and he only knows enough about a particular concept to solve a particular problem. The problem solving phase is extremely important because it normally leads to fruitful discussions. The success of these discussions is largely dependent on the maturity of the student's knowledge. This is the **U** phase or **UTILITY** phase. The student should be ready to use his knowledge. He should find pleasure in being confronted by questions and not feel threatened by it as normally happens. Drawing further on the bread comparison, this is the baking stage. Heat is added to turn the well-risen dough into a usable loaf of bread. Can the student stand the heat of problem solving? Unfortunately many students have not gone through the LEAVENING phase and the result is hardly usable.

Having gone through the three phases above, the student should be master of a particular topic. This is the level to which we would be happy to train our students. There is a fourth phase, the **O** phase. This is the phase where the student knows the topic so well that he can actually teach it himself. He can also evaluate different approaches to the same topic. This is the **OVERVIEW** phase. The student is so familiar with the concepts that he can view it from a distance. Not all students reach this level and it is not expected of most students either. In terms of our analogy, it is the phase where the loaf of bread can be shared around. This is of course the level of understanding that we expect from our tutors.

Hints on using this study guide

- Work with a friend or even in a group of three or four. It is much more fun and discussing a concept always makes you understand it better.

- A nursery school principle is that you learn through your hands. It still applies. Make the effort to fill in the blank spaces, even if you think you know the answer well. To convert your thoughts into writing is invaluable in the process of learning and securing concepts in your mind.

- Always keep a bird's-eye view of the subject. Look from above to see the whole picture and then orient yourself as to where you are in this picture at this moment.

- Most answers to the sections **Checking the concepts** and **Checking the examples** are included at the back. BEWARE! Solutions are the crutches of mathematics. You become so dependent on these that you have no confidence whatsoever without them. Refer sparingly to the answers.

- Work in PENCIL. You might want to change what you have written, or you might want to erase a section to repeat it before the exam.

Comment

This study guide was born from a sincere belief that students need more guidance in working through a calculus textbook packed with ideas.

A quote from Winnie the Pooh tells it all:

Owl explained about the Necessary Dorsal Muscles. He had explained this to Pooh and Christopher Robin before, and had been waiting ever since for a chance to do it again, because it is a thing which you can easily explain twice before anybody knows what you are talking about.

Less explaining and more doing!

Ansie Harding
aharding@scientia.up.ac.za

Contents

CHAPTER 1

FUNCTIONS AND CHANGE

1.1   WHAT IS A FUNCTION?

CAREFULLY READ THROUGH THIS SECTION IN THE TEXTBOOK

Key points

- Defining a function.

- Four ways of representing a function.

- Introducing functional notation.

Knowing the Book

1. A function is a .... that takes certain numbers as .... and assigns to each a definite .... number.

2. The set of .... is called the domain of a function.
   The set of .... is called the range of a function.

3. The input is also called the .... *variable* and the output is the .... *variable*.

4. List four ways in which functions can be represented:
   ....
   ....
   ....
   ....

5. Write "y is a function of t" in symbols: ....
   The dependent variable is .... and the independent variable is ....

Checking the concepts

1. Find the *three functions* amongst the following:

   (a) The number of skyscrapers decreases rapidly as you move away from the city center.

   (b) In 1950 the world population was 2.6 billion and in 1991 the world population was 5.4 billion.

(c) The car was driving too fast when it left the road.

(d) The more you have, the more you want.

(e) The summer has been particularly hot.

2. Can you draw graphs of the three functions that you chose above (marking the axes appropriately)?

3. Match the description of the function and the type of graph that would best describe it:

(1)	rainfall per year for the last five years	(a)	graph
(2)	love of classical music against age	(b)	table
(3)	the area of a circle against the radius	(c)	words
(4)	your change in moods during the day	(d)	formula

4. Write in symbols:

- B is a function of a: ....

- Given H as input, the output is Q: ....

- The value of P depends on the value of S: ....

Take note

Not all functions have smooth curves as graphs. The graph of a function given in a table is a set of points.

Checking examples

- **Example 1**: This example gives a practical example of how to use functional notation. Important here is to be able to interpret what is given in symbols. Note, for example, that the value of the car is given in *thousands* of dollars. So $f(3) = 6$ means that after 3 years the value of the car is 6000 dollars. This example also introduces the concept of a decreasing function. The value of the car *decreases* as time *increases*. It is of course not true that all decreasing functions are straight lines.

Problems

1, 2, 5, 11, 13 - 17

1.2   LINEAR FUNCTIONS

CAREFULLY READ THROUGH THIS SECTION IN THE TEXTBOOK

Key points

- Defining a linear function.

- Introducing the delta notation.

- Interpreting the slope of a linear function as a rate of change.

- Finding the equation of a line given the slope and one point on the line.

- Recognizing data from a linear function.

- Looking at families of linear functions.

Knowing the Book

1. Linear functions are functions whose graphs are ....

2. The slope of a linear function can be calculated from the values of the function at .... points, say $x = a$ and $x = c$, using the formula

$$\text{Slope} = \ldots\ldots\ldots\ldots = \ldots\ldots\ldots\ldots\ldots\ldots\ldots\ldots\ldots\ldots\ldots\ldots$$

The quantity .... is called the *difference quotient*.

3. A **linear function** has the form $y = $ .... Its graph is a line such that

- .... is the slope, or rate of change of y with respect to x.

- .... is the vertical intercept, or value of y when x is ....

4. The equation of a line with slope m through the point (x_0, y_0) is

....

5. How do you recognize that a table of x and y values comes from a linear function?

   ....

   ....

6. What do the lines $y = mx$, for different values of m, have in common?

   ....

   What do the lines $y = b + x$, for different values of b, have in common?

   ....

Checking the concepts

1. If a population P increased from 1000 in 3000 in 5 years, then $\Delta P = $ .... and $\Delta t = $ .....

2. If the temperature T at $t = 0$ was $20^0 C$ and $\Delta T = 30^0 C$ over $0 \leq t \leq 3$, then at $t = 3$, $T = $ ....

3. For the function f, the expression $\frac{f(4) - f(1)}{4 - 1}$ is called a .... quotient.

4. A line passing through the points (-1,2) and (3,6) has a slope of ....

5. For a function $s = f(r)$ the slope is given by $\frac{\Delta....}{\Delta....}$

6. Is the linear function $y = -0.5x + 3$ decreasing or increasing?

Checking examples

- **Example 1:** What would, according to this example, the world record time for the mile be in 2000?

- **Example 2:** What is the physical meaning of the value of the slope $m = 1.875$?

  ....

  Why is it reasonable to think that the true amount of waste in 2020 will be more than 261.45 million tons?

  ....

- **Example 3:** Note that in each of these tables there is an equal increase in the values of x. We therefore only need to check whether the increases in y are also constant.

- **Example 4:** Note that if q is a function of p, the slope is given by $\frac{\Delta q}{\Delta p}$; if p is a function of q, then the slope is given by $\frac{\Delta p}{\Delta q}$.

Problems

3, 10, 13, 19, 21, 29

1.3   RATES OF CHANGE

CAREFULLY READ THROUGH THIS SECTION IN THE TEXTBOOK

Key points

- Looking at average rate of change.

- Defining increasing and decreasing functions.

- Visualizing rates of change.

- Defining concavity.

- Looking at the relationship between distance and velocity.

Knowing the Book

1. For $y = f(t)$ the **Average rate of change** of y between $t = a$ and $t = b$ is given by:

$$\frac{\Delta y}{\Delta t} = \text{_____}$$

2. A function is increasing if ....
.... and decreasing if ....
....

3. Visualizing rate of change:
   Match the following:
   1. change   (a)   the slope of the line joining two points
   2. rate of change (b)   a vertical distance

4. A function is concave ...... if it bends downward as we move from left to right.

5. **Average velocity** is given by

$$\frac{\text{Change in } ....}{\text{Change in } ....}$$

6. Average velocity is the average .... of distance with respect to ....

Checking the concepts

1. Identify whether the underlined phrase or number refer to a rate of change:

   (a) Kari's height increased by <u>10 inches between the age of 5 and 10</u>.
   (change / rate of change)

   (b) Kari's height increased by <u>2 inches per year</u> between the age of 5 and 10.
   (change / rate of change)

   (c) By spending <u>$10000 more</u> on advertising the company gained an <u>additional $30000</u>
   in profits (change / rate of change)

   (d) Close to the summit of Mount Everest climbers cover approximately
   <u>140 feet per hour</u>. (change / rate of change)

2. Complete the following:

   - Example: Average growth rate (of a child)$= \frac{Change\ in\ height}{Change\ in\ time}$

   - Average cooling rate = ...

   - Average price increase = ...

3. A population of sheep decreased from 200 to 150 in five years, which means
   that the average rate of change over these five years is given by ....
   .... (include units). The population dropped
   by another 80 in another five years. During these five years the average rate of
   change ....-creased to .... Over the ten year period
   the average rate of change is given by ....

4. A linear function with a with a negative slope is ....-creasing.

5. A positive rate of change points to a function that is ......-creasing.

6. The path of a tennis ball across a court is concave ......

Checking examples

   - Example 1 makes use of a function defined in a table and Example 2 makes use
   of a function defined by a graph. In both cases we have enough information
   to calculate the average rate of change.

- **Example 3**: (*PCB's and Pelicans*):

  To make sure that you know what this example is all about:

  PCB is an industrial pollutant, and the concentration of PCB is measured in .... and the thickness of the egg shell in ....

  If the concentration of PCB increases from 204 ppm to 356 ppm, the thickness of an egg shell ....-creases from .... to .... (read from the table). This is a change of .... in the thickness of the egg shell for a change of .... in the concentration of PCB. The average rate of change then is .... (include units). This means that if the PCB level *in*creases by 1 unit, the thickness of the egg shell ....-creases by approximately .... mm.

- **Example 5** This example illustrates that a function can be increasing over a certain interval and decreasing over another. The same is true for being concave up or being concave down.

- **Example 6** This example makes use of a table and so the graph consists of points. Note the phrase "the graph appears to be concave up". We can only speculate about the concavity, assuming that the function does not do any funny turns between points.

Problems

1, 7, 11, 12, 13, 22

1.4   APPLICATIONS OF FUNCTIONS TO ECONOMICS

CAREFULLY READ THROUGH THIS SECTION IN THE TEXTBOOK

Key points

- Defining a

  - cost function

  - revenue function

  - profit function and a

  - depreciation function.

- Introducing marginal cost, marginal revenue and marginal profit.

- Introducing supply and demand curves.

- Looking at budget constraints.

Knowing the Book

1. The **cost function**, ..., gives the .... of producing a certain quantity .... of some good.

2. What is the difference between fixed cost and variable cost? ....

3. The *revenue function*, ..., represents the .... received by a firm from selling a quantity ...... of some good.

4. Revenue = ... × ....

5. For a linear cost function $C(q)$ the vertical intercept represents .... and the slope represents the ....

6. For a linear revenue function the slope of the line is ....

7. *Profit* is given by .... minus ...., so $\pi =$ ....

8. What is meant by the *break-even point* for a company? ....

9. What does the term *marginal cost* refer to? ....

10. What does the dependent variable of the *depreciation function* $V(t)$ of an item represent? ....
    What does the slope of the depreciation function tell you? ....

11. The *supply curve* represents how the quantity of an item that manufacturers are willing to .... per unit time depends on the ....
    The demand curve represents how the quantity of an item ....
    by consumers per unit time depends on the ....

12. What is meant by *equilibrium price* and *equilibrium quantity*?

13. What happens to the equilibrium price if specific tax is imposed on suppliers?

....

Checking the concepts

1. Of the following:

   (a) $C(q) = 2.3q + 17.5$

   (b) $C(q) = 10q^2 + 20$

   (c) $C(q) = 20(q + 5)$

   (d) $C(q) = \frac{10}{q} + 25$

   .... are typical cost functions.

2. If the fixed cost for production is $200 and the variable cost is $7 per item then the cost function is given by:..

3. Badges are sold for $7/item. The revenue function is given by:.......................

4. Of the following:

   (a) $R(q) = 10q + 15$

   (b) $R(q) = 20q$

   (c) $R(q) = 100 + 7q$

   (d) $R(q) = 0.1q$

   .... are typical revenue functions.

5. If the *depreciation function* of a machine is given by:

   $$V(t) = 10000 - 500t \qquad t \text{ in years and } V \text{ in dollars}$$

   then 10000 represents the ....and
   500 represents the ....
   The value after 5 years is ....

6. Decreasing or increasing?

   cost function   ....
   revenue function   ....
   depreciation function   ....
   supply function   ....
   demand function   ....

7. The marginal cost when 100 items are produced is $2000. What does that mean? ....

Checking examples

1. Read through Examples $1, 2, 3, 4, 5$ and 6. These are good for obtaining a firm grip on the many new concepts.

2. The Discussion on the Effect of Taxes on Equilibrium :

This discussion needs to be studied carefully. The outcome may seem surprising at first, but is extremely interesting. Repeat the example with a specific tax of $10 imposed per unit upon suppliers.

3. The Discussion on a Budget Constraint :

The idea of an implicitly defined function is discussed here. Any point lying within the triangle formed by the axes and the budget constraint function will be within the budget. Give an example of such a point.

Problems

1, 3, 6, 8, 11, 18, 23

1.5   EXPONENTIAL FUNCTIONS

CAREFULLY READ THROUGH THIS SECTION IN THE TEXTBOOK

Key points

- Defining an exponential function.

- Looking at examples of exponential growth and decay functions.

- Comparing linear and exponential functions.

- Recognizing data from an exponential function.

- Getting to know the family of exponential functions

- Introducing the number e.

Knowing the Book

1. P is an **exponential function** of t with base a if

   ....

   P_0 is the .... and a is the .... by which

   P changes when t ....

2. The factor a is given by $a = 1 + r$ where r is the ....

3. If a..........1 we have exponential *growth* and if a..........1 we have exponential *decay*.

4. Values of t and P could come from an exponential function if ratios of P are .... for equally spaced t values.

5. If $a > 1$ then a^t is ......-creasing and if $0 < a < 1$ then a^t is ....-creasing.

Checking the concepts

1. The function $A = 10(1.025)^t$ grows at ....% per time unit
   The function $A = 10(0.98)^t$ decays at ....% per time unit.

   (Remember that 2% means $\frac{2}{100}$ which is 0.02.)

2. For the exponential function $P = 25(1.20)^t$:
   the *initial value* (the value when $t = 0$) is ....
   If t increases from 0 to 1 then P will increase from .... to ....
   If t increases from 1 to 2 then P will increase from .... to ....

3. Two incomplete tables are given below. Complete the first table for an *exponential function* and the second table for a *linear function*:

x	1	2	3	4	5
y	1.2	1.8	....	....	....

x	1	2	3	4	5
y	1.2	1.8	....	....	....

4. If you start of with \$100 and it changes by a factor of 1.12 every year, then the amount of money you have at any time t (in years) is given by

   ....

5. True/False?

(a) Exponential growth at a rate of 3.5%, from an initial amount of 100 is given by $P = 100(1.35)^t$ T/F

(b) For any value of t, the function $P = 10(0.2)^t$ is positive and increasing. T/F

(c) The function $P = 100t^{1.35}$ is an exponential function.   T/F

(d) For exponential growth there is always a *constant growth factor*.   T/F

(e) An exponential growth function is concave up and an exponential decay function is concave down.   T/F

6. Match the function and the description:

1. $P = 10(1.3)^t$ (a)   the function with a 100% growth rate
2. $P = 10(0.8)^t$ (b)   the most rapidly decreasing function
3. $P = 20(2)^t$ (c)   the function for which $P(3) = 21.97$
4. $P = 10(0.4)^t$ (d)   the function with a decay rate of 20%

7. Use your calculator to find the following to 3 decimals:

- $e^2 = $ ....
- $e^{0.03} = $ ....
- $\sqrt{e} = $ ....
- $\frac{1}{e} = $ ....
- $\left(\frac{e}{8}\right)^{\frac{1}{3}} = $ ....

Checking examples

- The discussions on population growth and on elimination of a drug from the body form the backbone of this section. Study them well.

- **Example 1:** This is an important example for illustrating the fact that linear functions have constant absolute rate of change and exponential functions have constant relative (percent) rate of change.

- **Examples 2 and 3:** These examples are relatively simple but illustrate the underlying concepts well.

Problems

1, 3, 5, 7, 12, 13, 20

1.6   THE NATURAL LOGARITHM

CAREFULLY READ THROUGH THIS SECTION IN THE TEXTBOOK

Key points

- Defining the natural logarithm of x.

- Introducing exponential functions with base e

- Discussing the relationship between a^t and e^{kt}.

Knowing the Book

1. $\ln x = c$ means $x =$...........

2. Describe in words: $\ln x$ is the .... needed to get x. (Make sure that you know what this means.)

3. $\ln x$ is not defined if x is .... or ....

4. For $0 < x < 1$ the value of $\ln x$ is ....-tive For $x > 1$ the value of $\ln x$ is ....-tive

   The ln-function crosses the x-axis at $x =$ ....

5. $\ln e =$ ....

   $\ln 1 =$ ....

   $\ln e^x =$ ....

   $e^{\ln x} =$ ....

6. We know that e^x climbs extremely quickly. The natural logarithm function or $\ln x$, on the other hand, climbs extremely .... However, the ln-function does go to infinity as x increases.

7. Any **exponential** function can be written in either of two forms

   $$P = \text{......................................} \quad \text{or } P = \text{......................................}$$

8. To convert between a^t and e^{kt} we use $a = e^k$ so $k =$.................

9. Growth or decay?

   a^t with $0 < a < 1$ ....

   e^{kt} with $k > 0$ ....

   a^t with $a > 1$ ....

   e^{kt} with $k < 0$ ....

10. For e^{kt}, k is called the .... growth or decay rate.

11. If $a > 1$ and $a = e^k$ then k is ....-tive

    If $0 < a < 1$ and $a = e^k$ then k is ....-tive

Checking the concepts

1. This exercises will assist you in developing a feel for the functions e^x and $\ln x$.

   Keeping in mind that

   $e^0 = 1$

   $e^1 = 2.718282$

   $e^2 = 7.389056$ and

   $e^3 = 20.085537$ (all to 6 decimals), then *estimate* (very roughly) the missing numbers:

   - $e^{............} = 10$ (The number 2.1 would be a good estimate, for example)
   - $e^{............} = 2$
   - $e^{............} = 18$
   - $e^{............} = 4$
   - $e^{............} = 40$

   so that

   - $\ln 10 \approx$ ....
   - $\ln 2 \approx$ ....
   - $\ln 18 \approx$ ....
   - $\ln 4 \approx$ ....
   - $\ln 40 \approx$ ....

2. $\ln 50 = 3.912$ means that $e^{............} =$ ....

3. $e^{1.45} = 4.263$ means that $\ln$ .... $=$ ....

4. TEN TIMES: True or false?

   (a) $\ln e = 0$.   T/F

   (b) $\ln 102 = \ln 2 + \ln 100$.   T/F

   (c) $\ln 100 = 4.60517$ means that $e^{4.60517} = 100$.   T/F

(d) $e^{\ln 2.345} = 2.345$.   T/F

(e) $\frac{\ln 10}{\ln 3} = \ln 10 - \ln 3$.   T/F

(f) If $3.56 = e^{3t}$ then $t = 0.423254$.   T/F

(g) $\ln \frac{1}{x} = -\ln x$ T/F

(h) $\ln(5e^{3t}) = 5(3t)$.   T/F

(i) $\ln e^{4.567} = 4.567$.   T/F

(j) $e^3 \cdot e^4 = e^{12}$.   T/F

5. Convert the following:

(a) $(1.04)^t = (e^{\cdots\cdots\cdots})^t = e^{\cdots\cdots\cdots t}$

(b) $e^{0.055t} = (e^{0.055})^t = (\ldots\ldots\ldots)^t$

6. Match the following (without using a calculator):

(a) 2^t (i) $e^{-1.609t}$

(b) 6^t (ii) $e^{0.693t}$

(c) $(0.8)^t$ (iii) $e^{-0.223t}$

(d) $(0.2)^t$ (iv) $e^{1.792t}$

7. The function $P = e^{0.3t}$ shows a continuous growth rate of ....%

The function $P = e^{-0.05t}$ shows a continuous decay rate of ....%

8. The function $P = e^{0.05t}$ is ....-creasing and concave ....

The function $P = e^{-0.05t}$ is ....-creasing and concave ....

Take note

Remember that $\ln e^{whatever} = whatever$ and $e^{\ln whatever} = whatever$.

Checking examples

- **Examples 1-3:** In each of these examples the unknown variable t appears in the exponent. Therefore logarithms have to be used to find the solution. Expressions of the kind $\frac{\ln 2}{\ln 1.08}$ often appear. This is not the same as $\ln \frac{2}{1.08}$.

- **Example 4:** This example illustrates the conversion between e^{kt} and a^t.

Problems

1, 2, 8, 9, 20, 21, 22, 26 - 29, 38

1.7   EXPONENTIAL GROWTH AND DECAY

CAREFULLY READ THROUGH THIS SECTION IN THE TEXTBOOK

Key points

- Looking at doubling time and half-life of an exponential function.

- Looking at Financial applications: Compound interest and Present and Future value.

Knowing the Book

1. The **doubling time** of an exponentially ....-creasing quantity is the time required for the quantity to ....

2. The **half-life** of an exponentially ....-creasing quantity is the time required for the quantity to ....

3. If an amount of $\$P_0$ is deposited in an account paying interest at a rate of r (eg. $r = 0.06$) per year and P is the balance in the account after t years then

   $P = P_0$.................., if the interest is compounded annually and

   $P = P_0$.................., if the interest is compounded continuously.

4. If the continuous growth rate is 5%, will the annual growth rate be more or less than 5%? ....

5. The *future value B*, of a payment P, is the amount to which P ....
   ....

6. The *present value P*, of a future payment B, is the amount that would have to be ....

7. For interest compounded annually at rate r for t years:

   - Given the present value P then the future value $B =$ ....
   - Given the future value B then the present value $P =$ ....

8. For interest compounded continuously at rate r for t years:

   - Given the present value P then the future value $B =$ ....
   - Given the future value B then the present value $P =$ ....

Checking the concepts

1. A half-life of 20 years means that: A quantity of 1000 will decay to .... in .... years.

    The quantity of 500 will decay further to .... in another .... years.

2. Does the function $P = (0.6)^t$ have a constant doubling time or constant half-life? ....

    Does the function $P = e^{0.6t}$ have a constant doubling time or half-life?

    ....

3. True or false:

    - The function $P = 100(1.2)^t$ has a doubling time of $t = 3.8$.   T/F
    - Because $(1.1)^{7.27} = 2$ then $P = (1.1)^t$ has a doubling time of 7.27.   T/F
    - The functions $P = 20(1.02)^t$ and $P = 35(1.02)^t$ have the same doubling time.   T/F

4. What is the doubling time of $y = e^t$? ....

5. The expression

    $$P = 2000(1.08)^t$$

    describes the growth of an amount of ...., invested at ....% interest, compounded ....

    The expression

    $$P = 2000e^{0.08t}$$

    describes the growth of an amount of ...., invested at ....% interest, compounded ....

6. True or false: (Assume an 8% interest rate compounded continuously.)

    - The future value in 10 years of $100 is $222.55   T/F
    - The amount of $500, 10 years into the future, has a present value of $224.66   T/F

7. Where does the *rule of seventy* come from? (Keep in mind that $\ln 2 \simeq 0.7$.)

Take note

To calculate the doubling time of an exponential growth function $P = P_0 a^t$, set $a^t = 2$ and solve for t, so $t = \frac{\ln 2}{\ln a}$

To calculate the half-life of an exponential decay function $y = P_0 a^t$, set $a^t = 0.5$ and solve for t, so $t = \frac{\ln 0.5}{\ln a}$.

Equivalently: If $k > 0$, for the exponential growth function $P = e^{kt}$ the doubling time is $t = \frac{\ln 2}{k}$ and if $k < 0$, the exponential decay function $P = e^{kt}$ has a half-life of $t = \frac{\ln 0.5}{k}$.

Checking examples

- **Example 1**: Note the use of the ln-function in solving for t.

- **Example 2:** Note that information at two points is necessary to obtain values for P_0 and for k.

- **Example 3**: See the hint above.

- **Example 4**: Every exponential function has its own doubling time or half-life. Notice however that changing the initial value does not change the doubling time or half-life. In other words $P = 5000e^{-0.0025t}$ still has a half-life of 277 years.

- **Example 5**: The example illustrates the difference between annual and continuous compounding and continuous compounding is the winner.

- **Example 6**: A simple example on continuous compounding.

- **Example 7**: Note again that we need not know what the initial amount is, but by assuming it to be P_0, it will double to $2P_0$. The *rule of seventy* works for small percentages, but not so well for large percentages. You can show that for an annual growth rate of 25%, the actual doubling time is 3.1 years but according to the *rule of seventy* it is only 2.8 years.

- **Example 8**: This example illustrates that in order to compare the two circumstances you need to translate both to either the future or the present.

Problems

3, 5, 8, 10, 12, 22, 34

1.8   NEW FUNCTIONS FROM OLD

CAREFULLY READ THROUGH THIS SECTION IN THE TEXTBOOK

Key points

- Obtaining New Functions from Old through:

    - Composite functions.

    - Stretches of graphs.

    - Shifted graphs.

Knowing the Book

1. A composite function is a function of a .... such as $f(g(t))$ where we have an ....-side function g and an ....-side function f.

2. Multiplying by a negative constant c with $c < -1$ both stretches the graph and .... the graph about the x-axis.

3. The graph of $y = f(x)$ moved up k units (down if k is negative) is given by ....

4. The graph $y = f(x)$ moved to the right k units (to the left if k is negative) is given by ....

5. Multiplying by a constant $c > 1$ stretches the graph ....

Checking the concepts

- True or False?

    1. $f(x - 3) = f(x) - 3$ T/F

    2. $f(x+3)$ has the same graph as $f(x)$, only it is moved to the left.   T/F

    3. Multiplying a function by -0.5 shrinks the function and reflects it about the x-axis.   T/F

    4. If $f(t) = t^2$ and $g(t) = \sqrt{t}$, $t > 0$, then $f(g(t)) = 1$.   T/F

    5. The function $3 - f(x)$ is the function f shifted up 3 units and then reflected about the x-axis.   T/F

    6. $f(g(t)) = g(f(t))$ T/F

Checking examples

- **Example 1 - 4:** There are enough examples here to secure the concept of composite functions. Good preparation for later chapters.

- **Example 5:** Part (b) needs careful thinking.

Problems

2, 4, 7, 11, 12, 19, 21,

1.9   PROPORTIONALITY, POWER FUNCTIONS AND POLYNOMIALS

CAREFULLY READ THROUGH THIS SECTION IN THE TEXTBOOK

Key points

- Discussing proportionality.

- Defining a power function.

- Looking at graphs of power functions.

- Explaining what a polynomial is.

Knowing the Book

1. y is proportional to x if there is a nonzero constant k such that ....

2. $Q(x)$ is a **power function** of x if $Q(x)$ is proportional to a constant power of x. In symbols:

$$Q(x) = \text{.....................}$$

3. The formula for a polynomial is:

   ....

   where n is called the ....of the polynomial.

4. The term $a_n x^n$ is called the .... term and the non-zero number a_n is called the ....

5. An nth degree polynomial "turns around" at most .... times, but there may be .... turns.

Checking the concepts

1. If (a) $f(x) = 2^x$ (b) $f(x) = x^2$ and (c) $f(x) = x^{\frac{1}{2}}$
   then choose between (a), (b) or (c) in each of the following cases:
   1. The function that will dominate as x becomes large.   ....
   2. The function that is concave down.   ....
   3. The function that dominates between $x = 0$ and $x = 1$.   ....

2. Which of the following looks like $y = x^2$? ....

   (a) $y = x^{\frac{5}{3}}$

   (b) $y = x^{\frac{1}{2}}$

3. Which of the following functions are concave up for positive x?

   (a) $y = 23.5x^{1.45}$

   (b) $y = \frac{1}{x^6}$

   (c) $y = x^{0.3}$

4. Which is the *smaller* number in each case?

   - $10^{\frac{1}{3}}$ or $10^{\frac{1}{2}}$? ....
   - 2^x or x^2 if $x = 10$? ....

5. Write in symbols: y is inversely proportional to the square root of t.
   ....

6. True or False?

   (a) $y = \frac{1}{x} + 3$ is a polynomial of degree -1.   T/F

   (b) $y = x^{\frac{1}{2}} + 3x^{\frac{1}{4}}$ is a polynomial of degree $\frac{1}{2}$.   T/F

   (c) A power function is always a polynomial.   T/F

   (d) A quintic polynomial "turns around" at most five times.   T/F

   (e) $y = 10x^5 - x^3 + x$ has degree 5 and a leading coefficient of 10.   T/F

   (f) $y = x^n$, n a positive integer, either "turns around" once or not at all.
   T/F

   (g) If the leading coefficient of a polynomial is positive then $f(x) \to \infty$ as
   $x \to \infty$.   T/F

Checking examples

- **Example 1**: Describe the graph of $H = f(B)$.

- **Example 5**:
  Let $k = 4.8$, then $N = 4.8A^{\frac{1}{3}}$ for this example.
  An island of area $A = 1000$ square miles has .... species of birds.
  An island of area $A = 8000$ square miles has .... species of birds.
  Draw a graph of the number of birds against the area of the island.

- **Example 6** Note that demand and supply functions are often linear (polynomial of degree 1) which then means that the revenue function will be quadratic (polynomial of degree 2).

- **Example 7** This example opens up a whole new world. By adjusting the window of your calculator or computer you can create pictures that will confuse the uninformed. Take note of the comment written below Example 7.

Problems

1 - 4, 14 - 16, 18, 25, 29, 35, 39

1.10   PERIODIC FUNCTIONS

CAREFULLY READ THROUGH THIS SECTION IN THE TEXTBOOK

Key points

- Defining a periodic function.

- Looking at amplitude and period.

- Defining the sine and cosine functions.

- Studying the graph of $y = A \sin Bt + C$.

Knowing the Book

1. For periodic functions: If we know one .... we know the entire graph.

2. The **amplitude** is .... between the maximum and the minimum values.

3. The **period** is the .... needed for the oscillation to ....

4. The amplitude of the sine function is .... and the period is ....

5. For the function $y = A \sin Bt + C$, the amplitude is .... and the period is .... with C the ....

6. The "larger" B the .... the period.

Checking the concepts

1. A periodic function that oscillates between a minimum of 22 and a maximum of 29 has an amplitude of ....
   A periodic function that oscillates between a minimum of -14 and a maximum of 4 has an amplitude of ....

2. If $\sin t = 0.74570$ for $t = 2.3$ (radians), then the t value, one period later, for which $\sin t = 0.74570$ is ....

3. $y = 3 \sin 4t$ has a period of ....
   and $y = 2 \sin \frac{t}{4}$ has a period of ....

4. If $\sin(\frac{\pi}{3}) = 0.866$ then

   - $\sin(-\frac{\pi}{3}) =$ ....
   - $\sin(\frac{7\pi}{3}) =$ ....

5. The graph of $y = 100 \sin \frac{\pi}{6} t + 20$ has a period of .... and has a maximum value of .... and a minimum value of ....

6. Write in symbols:
   A sine function decreases from a high of 30 to a low of 20 and oscillates between these two values. It completes a full cycle in 6 days.

   ....

Take note

The sine and cosine functions oscillate around the **average** of the maximum and the minimum values and the amplitude is **half the difference**. A maximum value of 40 and a minimum of 20 will lead to the sine function: $30 + 10 \sin Bt$.

Checking examples

- **Example 1 & 2:** Notice that *not all* periodic functions are sine or cosine functions.

- **Examples 3 & 4:** Note in these examples that the x-axis can be marked in units of π as in Example 3 or in integers such in Example 4. Remember that π is just slightly more than 3. The sine-graph cuts the axis in Figure 1.99 just after 3, that is at π.

- **Example 6:** The choices made regarding sines or cosines for the graphs are not cast in stone. For the first graph we could have said that it looks like a cosine graph shifted to the right by 3π. For the second one we could have said it looks like a sine graph shifted right by 1 unit.

- **Example 7:** This problem is typical of a real-life situation modeled by a periodic function. Real-life situations do not always follow a sine graph perfectly but the sine graph does offer a good approximation.

Problems

2, 5, 9, 10, 11, 12, 16, 19, 29

REVIEW OF CHAPTER 1

1. A number of situations are described below. In each case:

   - Name the type of function that would best describe the situation.

   - Draw a graph that describes the situation, marking the axes appropriately.

   - Use symbols to describe the situation, using number values for parameters where possible.

(a) A population that numbered 4.3 million in 1940 grows at a continuous rate of 1.57% per year.

Type of function: ....

Graph:

Formula: ....

(b) The number of daylight hours vary between 10 and 14 hours during the year in a particular town.

Type of function: ....

Graph:

Formula: ....

(c) An amount of $350 dollars is invested at 5.5% compounded annually and the value after t years is required.

Type of function: ....

Graph:

Formula: ....

(d) House prices were increasing well but then took a dip in the recession after which they improved again.

Type of function: ....

Graph:

Formula: ....

(e) Temperature decreased at a constant rate from $100°F$ to $60°F$ in 3 hours.

Type of function: ....

Graph:

Formula: ....

2. The average rate of change in a population P between time t_0 and time t_1 is given by:

....

3. Give an example of an increasing function that changes concavity at $t = 0$? .

4. If the average rate of change in corn production between 1940 and 1950 is -4 tons/year and we know that 120 tons were produced in 1940, then the production in 1950 was ....

5. If (2.5, 100) is a point on a typical supply curve, it physically means that .....

....

....

6. Complete the table below for a linear function:

0.2	0.4	0.6	0.8
23.6	....	....	48.8

Complete the table below for an exponential function:

0.2	0.4	0.6	0.8
23.6	35.4	....	....

7. The point where the cost and the revenue functions intersect is called the .... and this is where the profit is

....

8. For a population function $P = 312(1.035)^t$, the number 312 stands for the .... and 1.035 is the factor by which P ....-creases if the time ....

The population above grows at an annual percentage of .... per year.

Review Problems

2, 5, 7, 9, 15, 19, 26, 30, 41, 46

FOCUS ON MODELING

FITTING FORMULAS TO DATA

CAREFULLY READ THROUGH THIS SECTION IN THE TEXTBOOK

Key points

- Defining Linear Regression.

- Explaining what "Best Fit" means.

- Interpreting the slope of the regression line.

- Understanding regression when the relationship is not linear.

Knowing the Book

1. Many formulas we use are approximations, often constructed from ....
....

2. The regression line can be used to make ....

3. Making a prediction between two points is called ....

4. If, instead, we make a prediction outside the interval given in the data, our estimate is called an .... In general, ...
is safer than ....

5. The line of best fit is called the .... line.

6. The slope of the regression line tells you the expected change in the ....
.... variable (y) given a .... change in the ....-
.... variable (x).

7. What is exponential regression? ....
....

Checking the concepts

1. If the independent variable in the table is a and the dependent variable is p:

a (in lb)	3	5.5	8
p (in dollars)	23.7	41.2	58.5

then of the following:

(a) $a = 7p + 2.5$

(b) $p = 2.5a + 7$

(c) $p = 7a + 2.5$

(d) $a = 2.5p + 7$

only .... could be the regression line for this set of data.

2. *Interpreting the slope:* The slope of the regression line that you chose above is ...., also written as $\frac{\cdots\cdots}{1}$. It is often useful to interpret a slope in terms of units. In this case the units are *dollars/lb*. This means that if *one more* .... is purchased, the .... would go up by $ ....

3. According to regression line (c) above, 4 lb cost .... and 11 lb cost ....

4. If the independent variable in the table is t and the dependent variable is a:

t (in sec)	2	4	6
a (in cm^3)	9.8	40.2	159.9

then of the following:

(a) $t = 2.5(2)^a$

(b) $a = 2.5t + 4.8$

(c) $a = 2.5(0.8)^t$

(d) $a = 2.5(2)^t$

only .... could be the exponential regression curve for this set of data.

5. What is the percentage growth rate of the function you chose in 4? ....

Checking examples

- **Example 1:**

  - Predict the total sales when $5500 is spent on advertising.
  - From the table we know that spending $4000 on advertising will generate $117000 in sales. Furthermore, the regression line has a slope of 16.5. So, if we spend another $1000 on advertising, then the sales will rise by approximately .... to ....

● **Example 2:**

    – Using the regression line and interpolation, the cost of producing 80 items can be estimated at ....

    – From the table we know that it costs \$742 to produce 100 items. (This includes fixed cost.) How much would it cost to produce 1 item more? (Use the slope of the regression line.)...

Problems

    1, 4, 5, 7, 10, 12

COMPOUND INTEREST AND THE NUMBER e

CAREFULLY READ THROUGH THIS SECTION IN THE TEXTBOOK

Key points

    ● Discussing effective annual yield.

    ● Looking at continuous compounding.

Knowing the Book

1. The effect of *interest earning interest* is called ....

2. *Compounded quarterly* means that interest is added .... times per year.

3. The .... tells you how much interest an investment really pays.

4. The *annual percentage rate*, is abbreviated to .... and is also called the .... rate or .... rate.

5. If a nominal rate of r is compounded n times a year, then with an initial deposit of \$P, the balance after t years will be

    ....

6. $\lim\limits_{n \to \infty} (1 + \dfrac{0.08}{n})^n = $ ....

7. If interest on an initial deposit of $\$P_0$ is compounded continuously at an annual rate of r, the balance after t years will be

....

Checking the concepts

1. If $100 grows to $106.2315 in one year the effective annual yield is ....

2. Complete:
$100(1 + \frac{0.05}{2})^2 = $ ....
$100(1 + \frac{0.05}{4})^4 = $ ....
$100(1 + \frac{0.05}{12})^{12} = $ ....

   The first expression tells you that an amount of ...., invested at a rate of ...., compounded .... is worth ....after .... year.

   The second expression tells you that an amount of ...., invested at a rate of ...., compounded .... is worth ....after .... year.

   The row of numbers, calculated above, approaches 105.1271096

   So, 5% continuously compounded interest effectively yields ....%

3. Because $e^{0.08} = 1.08328$ a continuously compounded nominal rate of 0.08 is the same as an effective rate of ....

Checking examples

- **Example 4:** An interesting question: Will bank X still be better if it offered a 7% annual rate compounded quarterly?

- **Example 6:** Compounding 1000 or 10,000 times a year cannot be much different to continuously compounding the interest. Show this by calculating $e^{0.07}$.

Problems

4, 5, 8, 9, 10

FOCUS ON THEORY

LIMITS TO INFINITY AND END BEHAVIOR

CAREFULLY READ THROUGH THIS SECTION IN THE TEXTBOOK

Key points

- Comparing power functions.

- Looking at limits to infinity.

- Comparing exponential and power functions.

Knowing the Book

1. For positive powers, high powers always dominate lower powers as $x \to$ ....

2. $\lim_{x \to \infty} f(x) = L$ means that ....
   ....

3. *Every* exponential growth function eventually dominates .... power function.

4. Every .... eventually dominates $\ln x$.

Checking the concepts

1. Of each of the following pairs, which will dominate eventually, (a) or (b)? (Sketches are sure to help.)

   (a) $y = x^4$ (b) $y = x^{\frac{1}{4}}$ ....

   (a) $y = x^4$ (b) $y = 500x^2$ ....

   (a) $y = 0.0000001x^3$ (b) $y = x^2$ ....

   (a) $y = 2^x$ (b) $y = x^{20}$ ....

   (a) $y = e^{0.001x}$ (b) $y = x^{10}$ ....

2. Determine the following:

   - $\lim_{x \to \infty} \ln x =$ ....

   - $\lim_{x \to \infty} -x^2 =$ ....

   - $\lim_{x \to \infty} e^{-x} =$ ....

3. The end behavior of $x^9 - 100x^5 + 200x$ is the same as the end behavior of
....

Checking examples

- **Examples 8 and 9**: Example 8 shows how to calculate limits to infinity and Example 9 shows how to use these limits to determine the end behavior of any polynomial.

Problems

1, 3, 8 - 10, 25, 26

CHAPTER 2

RATE OF CHANGE: THE

DERIVATIVE

2.1 INSTANTANEOUS RATE OF CHANGE

CAREFULLY READ THROUGH THIS SECTION IN THE TEXTBOOK

Key points

- Defining instantaneous velocity.

- Defining instantaneous rate of change.

- The derivative at a point.

- Visualizing rate of change: Slope of the graph and slope of the tangent line.

- Estimating the derivative of a function given numerically.

Knowing the Book

1. The **instantaneous velocity** of an object at time t is defined to be the limit of the .... velocity of the object over ....
....

2. The **instantaneous rate of change** of f at a is defined to be the limit of the .... rates of change of f over ....
....
NB: This is one of the most fundamental definitions in the book. Do you know exactly what it means?

3. Since average rate of change is a difference quotient, the instantaneous rate of change is the .... of difference quotients.

4. NB: The *derivative of f at a*, written ...., is defined to be the ....
....

5. The derivative of a function at point A is equal to

- The slope of ....
- The slope of ....

6. If we zoom in on the curve around a point the graph of the function looks like a .... We can estimate the derivative at a point by zooming in and estimating the slope.

7.   • If a table of values is provided, the instantaneous rate of change at a point can be estimated by finding the .... rate of change just to the right or left of the point.

• Note that normally a better estimate can be obtained by averaging the average rates of change to the right and to the left of the point.

Checking the concepts

1. Which three of the following indicate instantaneous rate of change?

   (a) At the beginning of 1996 the local library lent out books at a rate of 125 books per day.

   (b) Over the first 5 months of 1996, a total of 10000 books were lent out, which works out at 2000 books per month.

   (c) The size of land per person that is used for farming decreased at the beginning of 1900 at a rate of 0.012 acres/person.

   (d) Over the past few months the local cinema has featured, on average, three new films per month.

   (e) A function $f(t)$ describes population size in millions as a function of time t, measured in years. The slope of the graph of f, at $t = 1$, has a value of 2.64.

2. A small interval around $t = 1.5$ is $1.45 \leq t \leq 1.55$
   A smaller interval around $t = 1.5$ is ....
   An even smaller interval around $t = 1.5$ is ....
   An even smaller interval around $t = 1.5$ is ....

3. The length of the interval:
   $1.99 \leq t \leq 2.01$ is ....
   $1.45 \leq t \leq 1.55$ is ....
   $0.999 \leq t \leq 1.001$ is ....

4. A small interval to the right of 0, with 0 as the left endpoint is ....
   A small interval to the left of 0, with 0 as the right endpoint, is ....
   A small interval to the left of 2, with 2 as the right endpoint, is ....
   A small interval to the right of 2, with 2 as the left endpoint, is ....
   A small interval to the left of 1.4, with 1.4 as the right endpoint, is ....

5. The average rate of change of $f(x) = 3(0.4)^t$ over

   (a) $2 \le x \le 2.001$ is ....

   (b) $-0.001 \le x \le 0$ is ....

   Why are both answers above negative? ....

6. The average rate of change of $f(x) = \sqrt{x}$ over the interval is:

   $1.9 \le x \le 2.1$ is .... (6 dec)

   $1.99 \le x \le 2.01$ is .... (6 dec)

   $1.999 \le x \le 2.001$ is .... (6 dec)

   A fair estimate for $f'(2)$ is ....(5 dec)

7. Match the following:
   (1) $f'(1) = 3$ (a) f has a horizontal tangent at $x = 1$.
   (2) $f(2) = 7$ and $f(4) = 13$ (b)   If x increases by 0.02 then
                                                                y increases by approximately 0.05.
   (3) $f'(1) = 0$ (c) $f'(3)$ is approximately 3.
   (4) $f'(1) = 2.5$ (d) f is increasing around $x = 1$.

8.   • Graphically the *instantaneous rate of change* of a function at **a point** is
         the slope of the line .... to the curve at that point.

      • Graphically the *average rate of change* **over an interval** is the slope
         of the secant line .... the two points corre-
         sponding to the endpoints of the interval on the graph.

9. If $f(2.5) = 12.6$, $f(3) = 11.1$ and $f(3.5) = 8.3$ then $f'(3)$ appears to be
   ....-tive and a good estimate for $f'(3)$ is ....

   ....

   (This can be calculated in one of three ways.)

10. Use symbols only to describe:

    (a) The instantaneous rate of change of P at $t = 4$ is 2.34 ....

    (b) The line tangent to f at $t = 5$ is horizontal. ....

    (c) Near $t = 3$, a unit change in t would cause a change of 5 in S. ....

Take note

When calculating a derivative from a table of values: If the x-values are evenly spaced (eg 2, 4, 6) the average of the derivative to the right and to the left of the point is the same as the value to the right of the point minus the value to the left of the point divided by twice the difference in x-values. For example:

Given the table $\begin{array}{c|ccc} \text{x} & 0 & 2 & 4 \\ \hline \text{y} & 4 & 9 & 13 \end{array}$ then $f'(2) \simeq = \dfrac{1}{2}(\dfrac{13-9}{2} + \dfrac{9-4}{2}) = \dfrac{13-4}{4} = 2.25$

Checking examples

- **Example 1:** A valuable example. Check whether you have the necessary understanding by working through the following:

  For the function $Q(t) = 25(0.8)^t$:

  The average rate of change over the interval $4.5 \leq t \leq 5$ is

  ....

  The average rate of change over the interval $5 \leq t \leq 5.5$ is

  ....

  The instantaneous rate of change of f at $t = 5$ lies between .... and
  ...., so a good estimate would be ....

  The average rate of change over the interval $4.9 \leq t \leq 5$ is

  ....

  The average rate of change over the interval $5 \leq t \leq 5.1$ is

  ....

  The instantaneous rate of change of f at $t = 5$ lies between .... and
  ....

  So a better estimate would be ....

- **Example 2:** In this case the estimate is exact.

- **Example 3:** If the slope of the tangent line at a point is positive, the derivative at the point is positive.

- **Example 4:** Here we use average rates of change to the left and to the right of the point to find bounds for the derivative.

- **Example 5:** This example illustrates the difference in interpretation of the slope of a tangent line, the slope of a connecting line and a vertical distance.

• **Example 6:** This example illustrates how the instantaneous rate of change can be estimated from a table. Such a numerical estimate is often very useful.

Have a closer look at the answer at the solution to (a):

Between 1800 and 1970 the amount of crop land per person was decreasing at an average rate of 0.0157 acres per person per year. It can be interpreted as follows:

1 year after 1800 every person had approximately .... acres less land for farming (on average, because some people might have more and others less). It also means that 10 years after 1800 every person had approximately .... acres less land for farming (on average).

Problems

1, 2, 3, 4, 6, 11, 13, 16, 19

2.2   THE DERIVATIVE FUNCTION

CAREFULLY READ THROUGH THIS SECTION IN THE TEXTBOOK

Key points

- Defining the **derivative function**.

- Finding or estimating the derivative of a function given in one of three ways:

    - Graphically

    - Numerically

    - As a formula

Knowing the Book

1. The important point to notice is that for every x-value, there is a .... value of the derivative.

    If f is a function, the derivative is also a ....

2. For any function f, we define the **derivative function**, f', by

    .... = ...

3. Fill in one of the following: below/above/cuts

   When sketching a derivative function: At points where f has a large upward slope, the derivative is positive and so the derivative graph is .... the x-axis.

   On the other hand, at points where the function has a negative slope, the derivative graph lies .... the x-axis.

   Where the function has a zero slope the derivative function .... the x-axis.

Checking the concepts

1. For a function $f(x) = x^2$, match the following:

3	the derivative function
$2x$	the derivative of f at 2
6	the instantaneous rate of change of f at 3
4	the average rate of change of f over $0 \le x \le 3$

2. For the given points on the graph of f and say whether f' is positive/negative/zero and then estimate the value of f' (assuming that the scales on the two axes are the same). (A good idea for doing this is to use your ruler as a tangent to the graph at a given point and then estimate the slope of the "tangent").

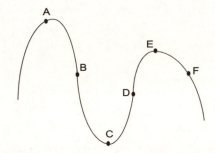

Point	f' pos/neg/zero	Value of f'	f' above/below/cuts x-axis
A	....	....	....
B	....	....	....
C	....	....	....
D	....	....	....
E	....	....	....
F	....	....	....

3. Match the following: (Making sketches should help a lot.)

(a) The derivative graph cuts the x-axis.

(b) The derivative graph drops even lower below the x-axis.

(c) The derivative graph is only just above the x-axis.

(d) The derivative graph is far above the x-axis.

1. The slope of the tangent to f is large and positive.

2. The value of the derivative decreases from -2 to -3.

3. The function changes from increasing to decreasing.

4. The tangent to f is almost parallel to the x-axis, but has a small positive slope.

Checking examples

- **Examples 1 & 2 :**
  These examples illustrate the principles for drawing a derivative function. Use your ruler and slide it along the graph of the function to see how the derivative changes. Spend a fair amount of time on these two examples. Once you feel comfortable about sketching a derivative graph, try the following: Sketch the derivative function of $f'(x)$. In other words, sketch the *derivative of the derivative*.

Problems

3, 6, 7, 15, 17, 29

2.3   INTERPRETATIONS OF THE DERIVATIVE

CAREFULLY READ THROUGH THIS SECTION IN THE TEXTBOOK

Key points

- Introducing an alternative notation for the derivative.

- Using units to interpret derivatives.

- Using the derivative to estimate values of a function.

Knowing the Book

1. An alternative notation for $f'(x)$ is introduced, namely, ..., also known as the notation of ....

2. The notation $\dfrac{dy}{dx}$ reminds us that the derivative is a limit of ratios of the form

   ....

   and reads "the derivative of .... with respect to ...."

3. The units of the derivative are units of the .... variable divided by the units of the .... variable.

4. The derivative is *approximately* equal to the change in the dependent variable when the independent variable increases by ....
   (if the function doesn't change rapidly near the point).

5. Local linear approximation means that if x increases by Δx then y will increase by

   $$\Delta y \simeq \text{.................................}$$

Checking the concepts

1. This paragraph illustrates a fundamental way of interpreting a derivative. Make sure that you grasp it fully:

   If, for a function $f(t)$, we know that $f'(4) = 6$, it means that if the t value increases by **1 unit**, the y-value increases by approximately **6 units**.

   So, if $f'(5) = 1.6$ it means that ....
   ....
   ....

   And if $P'(10) = -2.5$ it means that ....
   ....
   ....

2. Make sure that you grasp the concepts of local linear approximation:
   Say we know both the derivative and the function value at a point, for example: $f'(3) = 4$ and $f(3) = 2$. Then an approximation for the function value at a nearby point, say at $x = 3.5$ can be obtained by using

   $$y = 2 + \Delta y \simeq 2 + \Delta x f'(x)$$

In this case $\Delta x = 0.5$ and $\Delta y \simeq 0.5 \times 4 = 2$
So $f(3.5) = f(3) + \Delta y \simeq 2 + 2 = 4$.

If, for another function $f(t)$, we know that $f'(5) = 1.6$ and $f(5) = 2$ then an approximation for $f(7)$ is ....
....

If $P'(10) = -2.5$ and $P(10) = 5$ then an approximation for $P(12)$ is ....
....

Checking examples

- **Example 2**: The cost of building successively larger houses normally increases at a decreasing rate. This means that although $f'(A)$ is always positive, it decreases as A increases. Make a sketch of cost against area:

  Typical numbers are: $f(1000) = 60000$ and $f'(1000) = 50$ Physically this means that a house of .... square feet costs $ .... To build an additional square foot would cost approximately $ ...., so a house of 1200 square feet will cost ....

- **Example 3**: The physical meaning of $f'(2500) = \$90/\text{ton}$ is ....
  ....

- **Example 4**: If $f'(3) = 50$ as given, and in addition $f'(3.50) = 45$, what does it tell you about the shape of $q = f(p)$ around $p = 3$?
  ....
  ....

- **Example 5**: Study this example carefully. It illustrates the interpretation of both the function value and the derivative well.
  If $f(5) = 2$ it means that ....
  ....

If $f'(5) = 1$ it means that ....

....

Write the following phrases in symbols:

If 8 mg is administered the drug will stay in the body for 3.5 hours:

....

If the dosage is increased from 8mg to 9 mg the drug will stay in the body approximately 1.2 hours longer: ....

Problems

1, 2, 3, 8, 10, 11, 16, 17

2.4   THE SECOND DERIVATIVE

CAREFULLY READ THROUGH THIS SECTION IN THE TEXTBOOK

Key points

- Looking at what the second derivative tells us.

- Interpreting the second derivative as a rate of change.

Knowing the Book

1. If $f' > 0$ on an interval, then .... is increasing on that interval.
   If $f'' > 0$ on an interval, then .... is increasing on that interval, so that the graph of f is .... on that interval.

2. If the first derivative is the rate of change, the second derivative is the rate of change of ....
   If the second derivative is positive, the rate of change (the first derivative) is ....-creasing.
   If the second derivative is ...., the rate of change is decreasing.

3. The phrase : "Congress had merely cut the rate at which the defense budget was increasing" appears towards the top of p 146. Draw a graph of the shape

of the defense budget against time to explain what is meant by this statement.

Checking the concepts

1. Sketch a curve in each of the four cases below:
   (a) f' is positive and increasing (b) f' is negative and increasing (becoming less negative)

   Both these curves are concave .... because f' is ....
   (c) f' is positive and decreasing. (d) f' is negative and decreasing (becoming more negative).

   Both these curves are concave .... because f' is ....

2. For which of the four curves above is f'' positive? ....
   For which of the four curves above is f'' negative? ....

3. Sketch a curve for which f'' changes from positive to negative.

4. If for an increasing function f we know the following:

   $f(0) = 10.5$
   $f(2) = 12.8$
   $f(4) = 15.6$
   $f(6) = 18.9$

   then fair approximations for the derivative at various points are:

   $f'(0) = 1.15$
   $f'(2) = 1.275$
   $f'(4) =$ ....

   It is clear that f' is ....-creasing and therefore the curve of f is concave .... and f'' .... 0.

5. For $f'(0)$, $f'(2)$ and $f'(4)$ as above, then an estimate for $f''(2)$ is ....

Checking examples

- **Example 1:** The second derivative could become zero at a point but we cannot conclude anything regarding the concavity of the function at that point.

- **Example 2:** This example shows a population growth for which the second derivative changes from positive to negative at t^*, called logistic growth (discussed in Chapter 5). The growth rate is maximal at t^*. Can you sketch a population growth pattern for which the second derivative changes from negative to positive at t^*? What can you say about the rate of change at t^*?

- **Example 3:** Study the discussion on the sign of the second derivative (the (b) part) carefully.

  Which of the following is true:

  1. The number of abortions reported is increasing at an increasing rate.

  2. The number of abortions reported is increasing at a decreasing rate.

Problems

1, 2, 4, 7 - 12, 15, 17, 18

2.5   MARGINAL COST AND REVENUE

CAREFULLY READ THROUGH THIS SECTION IN THE TEXTBOOK

Key points

- Looking at graphs of cost and revenue functions.

- Doing marginal analysis.

- Maximizing Profit.

Knowing the Book

1. What is meant by *economy of scale*? ....
   ....

2. When an economic decision has to be taken, a crucial question is whether *additional costs* incurred are .... the *additional revenue* generated.

3. These additional costs and revenues are called .... costs and revenues.

4. The instantaneous rate of change of cost with respect to quantity is called the ....

5. If the price, p, is constant, then the revenue function is $R =$ .... and the graph of R against q is a .... through .... with slope equal to ....

6. Marginal cost is defined as MC = ...

   Marginal revenue is defined as MR = ...

7. Maximum profit occurs when ....

Checking the concepts

1. (a) If the units are dollars and q is the number of units produced then:
       If $C(120) = 1000$ and $R(120) = 1500$ the profit for manufacturing 120 items is ....

    (b) If $C'(120) = 10$ it means that it will cost **approximately** .... more to manufacture **one** more item (the 121st item).
       If $R'(120) = 18$ it means that if **one** item more than .... is sold, the revenue will increase by **approximately** ....
       Will the profit increase if more items are manufactured? ....
       By how much (approx) will the profit increase/decrease if one more item is manufactured? ....

    (c) If, on the other hand, $C'(120) = 15$ and $R'(120) = 10$ it means that if **one** extra item is manufactured (the 121st item) then it will cost approximately .... more and the revenue will increase by approximately .... The profit will .... -crease.
       So it will/will not be wise to produce more items.

    (d) If $C'(120) = R'(120) = 12$ it means that ....
       ....

2. If $MR > MC$ then .... (more/less) items should be manufactured.
   If $MR < MC$ then .... (more/less) items should be manufactured.
   If $MR = MC$ (and cost is less than revenue) then .... items should be manufactured.

3. For a linear revenue (or cost) function, the marginal revenue (or cost) is always

....

Checking examples

- **Examples 1** : What happens for quantities less than 130?

- **Example 3**: Remember that the slope of a curve at a point is also the slope of the tangent at the point. Remember to use a ruler as a tangent to judge the value of the slope.

- **Example 4**: An example to spend some time on. It may be useful to refer back to section 2.3 where derivative functions were sketched. The slope of the cost function is always positive and so the derivative function must lie above the q-axis. The slope of the cost function changes from large positive to small positive to large positive. So the derivative graph is far above the x-axis, low above the x-axis and far above again. Hence the parabolic shape.

- **Example 5**: Check your understanding of this important example: The profit is .... (negative/positive) at first. Then it is .... but ....(small/big). It then increases to a .... after which it .... again.

Problems

1, 2, 4, 5, 8, 9

REVIEW OF CHAPTER 2

1. A small interval around $t = 1.25$ is ....

   A smaller interval around $t = 1.25$ is ....

2. The **instantaneous rate of change** of f at a is given by the ....
   .... rates of change of f over ....
   ....

3. The average rate of change of $Q(t) = t^{2.4}$ over $0.9 \leq t \leq 1.1$ is:

   ....

   The average rate of change of $Q(t) = t^{2.4}$ over $0.99 \leq t \leq 1.01$ is:

The instantaneous rate of change of Q at $t = 1$ is approximately ....(2 dec)

4. For a function f, the following expressions can each be interpreted graphically:

   - $f'(1)$:

     The slope of ....

   - $\frac{f(3)-f(1)}{3-1}$:

     The slope of ....

5. If the graph of the *derivative function* is above the x-axis, then the *function* is

   ....

   If the graph of the *derivative function* cuts the x-axis then the function

   ....

6. For $N = f(s)$ the expression $N'(12) = 2.3$ means that if ....

   .... then ....

7. If f' is *decreasing* over an interval the f is concave ....

8. An example of a function where the second derivative changes from negative to positive and then again to negative is:

9. The marginal cost MC is the instantaneous rate of change of cost with respect to .... So $C'(10) = 25$ for a shoe manufacturing company means that ....

   ....

Review Problems

2, 9, 10, 13, 16, 19, 23, 24

FOCUS ON THEORY

LIMITS, CONTINUITY, AND THE DEFINITION OF THE DERIVATIVE

CAREFULLY READ THROUGH THIS SECTION IN THE TEXTBOOK

Key points

- Using the definition of a derivative.

- Discussing the idea of a limit.

- Discussing continuity.

- Using the definition to calculate the derivative.

Knowing the Book

1. What is the formula for the average rate of change of f between a and $a+h$?

   ....

2. The derivative function $f'(x)$ is defined as:

   ....

3. What does it mean if $lim_{x \to c} f(x) = L$? ....
   ....

4. A continuous function has no breaks, ... or ....

5. Numerically, continuity means that small errors in the independent variable lead to small errors in ....

6. The function f is continuous at $x = c$ if f is .... at $x = c$ and

   ....

7. When is a function continuous on an interval $[a, b]$? ....
   ....

8. Continuity demands that the behavior .... the point be consistent with the behavior .... the point.

Checking the concepts

1. Find the average rate of change of $f(x) = x^2$ between $x = 2$ and $x = 2 + h$.

   ....

   Find $f'(2)$ by letting h approach 0.

   ....

2. Why do we say the limit as h approaches zero of $\dfrac{2 + h}{h}$ does not exist? ....

   ....

3. Is $f(x) = \dfrac{1}{x - 1}$ continuous on $[0, 2]$? ....

4. For the function in Figure 2.70:

   If x approaches 1 from the right-hand side, what value does f approach? .... What is the value of $f(1)$? ....

Checking examples

- **Example 1**: Notice that the limit *predicts* what will happen *at* the point by looking at what happens *near* the point.

- **Example 2**: This is an example of where the function is not defined at the point $x = 0$. Yet by looking at what happens near the point we expect a value of 0.693 at the point.

- **Examples 3 & 4**: These examples show the difference between estimating a limit numerically (an approximation) and finding a limit algebraically (an exact value).

- **Examples 5 & 6**: These examples may seem long-winded but this is the only valid way of establishing a derivative function.

Problems

1, 3, 6, 7, 9, 12, 16, 18, 19, 22, 25

CHAPTER 3

SHORT-CUTS TO

DIFFERENTIATION

3.1 DERIVATIVE FORMULAS FOR POWERS AND POLYNOMIALS

CAREFULLY READ THROUGH THIS SECTION IN THE TEXTBOOK

Key points

The derivatives of:

- a constant function.

- a linear function.

- a constant multiple.

- a sum and difference.

- powers of x.

- polynomials.

Knowing the Book

1. If $f(x) = k$, then $f'(x) = $ ....

2. If $f(x) = b + mx$ then $f'(x) = $ ....

3. $\dfrac{d}{dx}[cf(x)] = $ ....

4. $\dfrac{d}{dx}[f(x) + g(x)] = $ ....

5. $\dfrac{d}{dx}(x^n) = $ ....

Checking the concepts

1. If f is stretched vertically by a factor of 3 to become a new function g and $f'(3) = 2$ then $g'(3) = $ ...... This means that if x increases by 1 (from 3 to 4) then $f(x)$ will increase by .... (approx) and $g(x)$ will increase by ....(approx)

2. If f is flipped over the x-axis to become a new function g and $f'(3) = -6$ then $g'(3) = $ .... This means that if x increases by 1 (from 3 to 4) the $f(x)$ will .... by .... (approx) and $g(x)$ will .... by .... (approx)

3. If f is shrunk to $f/2$, the slopes are all smaller by a factor of ....

4. Find the wrong answers and correct them, if possible, by only using the rules in this section:

   (a) $\dfrac{d}{dx}(2)^5 = 5(2)^4$

   (b) $\dfrac{d}{dx}x^{1.345} = 1.345x^{0.345}$

   (c) $\dfrac{d}{dx}3^x = x3^{x-1}$

   (d) $\dfrac{d}{dx}\left(\dfrac{1}{x^{\frac{1}{2}}}\right) = \dfrac{1}{\frac{1}{2}x^{-\frac{1}{2}}}$

5. If the function $y = x^2$ is flipped over the x-axis to become $g(x) = -x^2$, the sign of the first derivative is reversed. Is the sign of the second derivative also reversed? ....

6. The sum rule is often informally described as:
   "The derivative of a sum is the sum of derivatives".
   In similar fashion: The derivative of a difference is the ....
   of ....

   Also: The derivative of a constant times a function, is the ....
   times ....

Checking examples

- **Example 5:** This examples brings a number of concepts together. Make sure you grasp it.
  Show that the equation of the tangent line at $x = -2$ is $y = -x + 15$.

- **Example 8:** Describe the behavior of the second derivative of this cubic similar to the way in which the behavior of the first derivative is described.

Problems

1-21, 29, 35, 36, 41

3.2 EXPONENTIAL AND LOGARITHMIC FUNCTIONS

CAREFULLY READ THROUGH THIS SECTION IN THE TEXTBOOK

Key points

The derivatives of:

- e^x

- a^x, where a is a constant

- $\ln x$

Knowing the Book

1. $\dfrac{d}{dx}(e^x) =$ ....

2. $\dfrac{d}{dx}(a^x) =$ ....

3. $\dfrac{d}{dx}(\ln x) =$ ....

4. The graph of the *derivative* of $f(x) = 2^x$ lies .... (above/below) the graph of $f(x)$ and the graph of the derivative of $g(x) = 3^x$ lies .... the graph of $g(x)$. The function $f(x) =$ .... has the remarkable property that the derivative matches the original function.

5. What is the only value for a such that $\dfrac{d}{dx}a^x = a^x$? ....

6. The logarithmic function increases everywhere (for $x > 0$), so we expect the derivative to be ....-tive The logarithmic function is concave down everywhere and so we expect the derivative to be ....-creasing for all $x > 0$.

7. The slope of the logarithmic function is very .... near $x = 0$ and is very .... for large x. So we expect the derivative to tend to .... for x near 0 and to tend to .... for large x.

Checking the concepts

1. Which function are we talking about?

   (a) The rate at which the function changes at any point is given by the y-value in the point.

   ....

   (b) The rate at which the function changes at any point is given by the reciprocal of the x-value in the point.

   ....

2. The graph of the derivative of $f(x) = 1.5^x$ lies .... the graph of $f(x)$.
   The graph of the derivative of $f(x) = 2.5^x$ lies .... the graph of $f(x)$.
   The graph of the derivative of $f(x) = 3.5^x$ lies .... the graph of $f(x)$.

3. We know that $\frac{d}{dx}(\frac{1}{2})^x = -0.69314(\frac{1}{2})^x$. How can the negative sign be explained graphically?

   ....

4. Find the wrong answers and correct them:

   (a) $\frac{d}{dx} 2^{1.035} = 2^{1.035} \ln 2$ ....

   (b) $\frac{d}{dx}(2.3)^x = x(2.3)^{x-1}$ ....

   (c) $\frac{d}{dx} \ln 5.2 = \frac{1}{5.2}$ ....

5. The function $f(x) = e^x$ changes at a(n) ....-creasing rate, whereas the function $f(x) = \ln x$ changes at a(n) ....-creasing rate.

6. The function $f(x) = 3^x$ is increasing; its derivative is ....-creasing and lies .... the x-axis.

   The function $f(x) = (0.3)^x$ is ....-creasing; its derivative is ....-creasing and lies .... the x-axis.

Checking examples

- **Example 3:** Differentiation is applied to a real-life problem. Check your understanding by using the derivative of P to find the rate at which the population will increase at the start of 2010. Give your answer in millions of people per year as well as in number of people per day.

- **Example 4:** It is an interesting exercise to find the equation of the tangent line at $x = 0.01$ (a small x-value). Notice the large slope.

Problems

1-20, 25, 30, 35

3.3   THE CHAIN RULE

CAREFULLY READ THROUGH THIS SECTION IN THE TEXTBOOK

Key points

- Recognizing a composition of functions.

- Finding the derivative of a composition of functions.

Knowing the Book

1. Suppose y is a composite function of t, then we can write $y = f(z)$ with $z =$ .... for some inside function g and outside function ....
   According to the chain rule:
   $$\frac{dy}{dt} = \text{..............................}$$

2. If z is a function of t then

$$\frac{d}{dt}(z^n) = \text{..............................}$$

$$\frac{d}{dt}(e^z) = \text{..............................}$$

$$\frac{d}{dt}(\ln z) = \text{..............................}$$

$$\frac{d}{dt}e^{kt} = \text{..............................}$$

Checking the concepts

1. Complete:

   (a) $\dfrac{d}{dt}(z(t))^5 = $ ....

   (b) $\dfrac{d}{dt}\sqrt{(f(t))} = $ ....

   (c) $\dfrac{d}{dt}f(z(t)) = $ ....

   (d) $\dfrac{d}{dt}y(........) = \dfrac{dy}{dz}\dfrac{dz}{dt}$

   (e) $\dfrac{d}{dt}v(............) = ...\dfrac{dz}{dt}$

2. Find the wrong answers and correct them:

   (a) $\dfrac{d}{dx}e^{-0.01x} = e^{-0.01x}$

   (b) $\dfrac{d}{dx}(x^3 + x)^4 = 4(3x^2 + 1)^3$

   (c) $\dfrac{d}{dx}(1 + \frac{x}{10})^{10} = 10(1 + \frac{x}{10})^9$

   (d) $\dfrac{d}{dt}(1 - t)^4 = 4(1 - t)^3$

3. The derivative of $f(x) = e^{z(x)}$ is given by: ....

Take note

When using the chain rule, start from the outside and work your way inside. Keep multiplying as you keep differentiating.

Checking examples

- **Example 1:** This example serves to explain why the chain rule is defined the way it is.

- **Example 5:** Don't forget the minus sign when differentiating $e^{-0.2x}$.

- **Example 6:** How would the problem change if the interest was compounded annually?

Problems

2 - 30 (even numbers), 34, 36, 38

3.4 THE PRODUCT AND QUOTIENT RULES

CAREFULLY READ THROUGH THIS SECTION IN THE TEXTBOOK

Key points

- The product rule and

- the quotient rule.

Knowing the Book

1. Two ways of writing down the product rule are:

$$(fg)' = \text{..}$$

$$\frac{d}{dx}(uv) = \text{..}$$

2. Express the product rule in words.

3. Two ways of writing down the quotient rule are:

$$\left(\frac{f}{g}\right)' = \text{..}$$

$$\frac{d}{dx}\left(\frac{u}{v}\right) = \text{..}$$

4. Express the quotient rule in words.

Checking the concepts

1. By using the "constant times a function rule" we see that the derivative of $5x^2$ is $10x$. Will you get the same answer if you apply the *product rule* to $5x^2$?

   ....

2. According to the product rule:

   (a) $\dfrac{d}{dt}v(t)w(t) = \text{..}$

   (b) $\dfrac{d}{dt}s(t)s(t) = \text{..}$

   (c) $\dfrac{d}{dx}xf(x) = \text{..}$

(d) $\dfrac{d}{dp}p^2.f(p) =$ ....

3. According to the quotient rule:

(a) $\dfrac{d}{dt}\dfrac{v(t)}{w(t)} =$ ....

(b) $\dfrac{d}{dt}\dfrac{1}{s(t)} =$ ....

(c) $\dfrac{d}{dt}\dfrac{x}{f(x)} =$ ....

Checking examples

- **Example 3:** This example makes use of an exponential demand function rather than a linear demand function. Make sure that you follow it.

  Find the marginal revenue at $q = 21$ and at $q = 693$ and interpret your answers.

Problems

1, 3, 6, 9, 11, 22 - 26, 33, 36, 38

3.5 DERIVATIVES OF PERIODIC FUNCTIONS

CAREFULLY READ THROUGH THIS SECTION IN THE TEXTBOOK

Key point

- Derivatives of the sine and cosine functions.

Knowing the Book

1. $\dfrac{d}{dx}(\sin x) =$ .... and $\dfrac{d}{dx}(\cos x) =$ ....

2. If u is a differentiable function of t then

   $\dfrac{d}{dt}(\sin u) =$ .... and $\dfrac{d}{dt}(\cos u) =$ ....

3. Because $f(x) = \sin x$ increases on $-\frac{\pi}{2} < x < \frac{\pi}{2}$ the derivative function $f(x) = \cos x$ is ....-tive and therefore lies .... the x-axis on this interval.

4. The value of the *derivative* of the sine function at a point is the same as the value of the .... function at the same point.

Checking the concepts

1. If $\sin(\frac{\pi}{4}) = 0.707$ and $\cos(\frac{\pi}{4}) = 0.707$, then the sine graph has a slope of .... at $t = \frac{\pi}{4}$ and the cosine graph has a slope of .... at $t = \frac{\pi}{4}$.

2. If the sine graph has a slope of -0.6956 at $t = 2.34$, then the next value at which sine graph will have a slope of -0.6956 is ....

3. The derivative of the cosine function is zero at the points where the sine graph
....

4. If $\sin(3.4\pi) = -0.951$ then the cosine graph, at $t = 3.4\pi$ is ....-creasing.

Checking examples

- **Example 1:** Notice that $f'(0) = 1$ for both $f(x) = \sin x$ and $f(x) = x$. Furthermore, for both these functions $f(0) = 0$. Therefore, close to the origin, it is difficult to distinguish between $f(x) = \sin x$ and $f(x) = x$. This means that $\sin x \simeq x$ for small values of x.

  For example: $\sin(0.12) \simeq 0.12$ (Actual value to 4 decimal places is 0.1197)

Problems

1-16, 18, 20

REVIEW OF CHAPTER 3

1. Differentiate the following for a positive number n:

- $\dfrac{d}{dx}x^n =$ ....

- $\dfrac{d}{dx}x^{-n} =$ ....

- $\dfrac{d}{dx}\sqrt{x^n} =$ ....

- $\dfrac{d}{dx}x^{\frac{1}{n}} =$ ....

2. We know that $\dfrac{d}{dx}a^x = a^x \ln a$ (a positive). For what values of a is the derivative

- positive? ....

- negative? ....

3. (a) The ln function increases at a decreasing rate, therefore the derivative is
    ....-tive and ....-creasing.

    (b) The e-function increases at an increasing rate, therefore the derivative is
    ....-tive and ....-creasing.

4. If u is a function of t then

   • $\dfrac{d}{dt}(u(t))^{-2} =$ ....

   • $\dfrac{d}{dt}\sin u(t) =$ ....

   • $\dfrac{d}{dt}(u(t))^{6} =$ ....

   • $\dfrac{d}{dt}\ln u(t) =$ ....

5. If the demand function is $p = 20e^{-0.02q}$ and the revenue function is the product
   of quantity and price, then the revenue as a function of quantity is

   ....

   and

   $R'(q) =$ ....

6. Use both the quotient and the chain rules to show that $\dfrac{d}{dx}(\dfrac{1}{x}) = \dfrac{-1}{x^2}$.

Review Problems

7, 8, 9, 10, 23, 26, 29, 39, 44, 45

FOCUS ON THEORY

ESTABLISHING DERIVATIVE FORMULAS

CAREFULLY READ THROUGH THIS SECTION IN THE TEXTBOOK

Key points

- Using the chain rule to establish derivative formulas

- Using the definition of the derivative to prove:

  - The Product Rule.
  - The Quotient Rule.

Knowing the Book

For any function f, we define the **derivative function f'** by

....

Checking the concepts

1. (a) If $f(x) = 3$ then $f(x + h) = $ ....

   (b) If $f(x) = e^{3x}$ then $f(x + h) = $ ....

   (c) If $f(x) = x^2 + x$ then $f(x + h) = $ ....

   (d) If $f(x) = \dfrac{1}{x}$ then $f(x + h) = $ ....

2. In order to *prove* that the derivative of x^2 is $2x$, we have to make use of the

   ....

3. (a) $\displaystyle\lim_{h \to 0} \frac{1}{x + h} = $ ....

   (b) $\displaystyle\lim_{h \to 0} 3h = $ ....

   (c) $\displaystyle\lim_{h \to 0} (x^2 + xh + h^2) = $ ....

   (d) $\displaystyle\lim_{h \to 0} \frac{xh + h^2}{h} = $ ....

   (e) $\displaystyle\lim_{h \to 0} \frac{e^h - 1}{h} = $ ....

   (f) $\displaystyle\lim_{h \to 0} \frac{e^{3h} - 1}{h} = $ ....

4. If $f(x) = e^{3x}$ then $\dfrac{f(x+h) - f(x)}{h} = $ .... which simplifies to

....

Taking the limit as h approaches 0 (graphically or numerically) gives $f'(x) = $
....

Checking examples

- **Example 3:** If you don't have the use of a graphics calculator and you have to answer the question: "What is the limit of $\frac{e^h - 1}{h}$ as h approaches zero?," you can substitute smaller and smaller values of h into the expression.

   For example, for $h = 0.01$, $\dfrac{e^h - 1}{h} = $ ....

   For $h = 0.0001$, $\dfrac{e^h - 1}{h} = $ ....
   From this one can deduce that the expression probably approaches 1 as h approaches zero.

Problems

5, 6, 8, 11

FOCUS ON PRACTICE

DIFFERENTIATION

The purpose of this section is to practice, practice, practice. It is necessary to do many of these exercises. No other way out.

CHAPTER 4

USING THE DERIVATIVE

4.1 LOCAL MAXIMA AND MINIMA

CAREFULLY READ THROUGH THIS SECTION IN THE TEXTBOOK

Key points

- What the derivative tells us about its graph.

- Defining a local minimum and a local maximum.

- Defining a critical point and a critical value.

- Testing for local Maxima and local Minima: First Derivative Test and Second Derivative Test.

Knowing the Book

1. Suppose p is a point in the domain of f. Then

   - f has a local minimum at p if $f(p)$ is ....
   ....

   - f has a local maximum at p if $f(p)$ is ....
   ....

2. For any function f, a point p in the .... of f is a **critical point** if $f'(p) =$ .... or $f'(p)$ is ....
   In addition, the point .... on the graph is also called a critical point.
   A **critical value** of f is the value .... at ....

3. Between two successive critical points (assuming the function is defined for all points in between) the graph of a function cannot change direction; it either .... or ....

4. If a function has a local maximum or minimum at p then p is either a ....
   .... or an .... of the interval.

5. You suspect that f has a local maximum at a critical point p where $f'(p) = 0$.
   How do you test whether this is true or not by using the *first* derivative test?
   ....
   ....

How do you test whether this is true or not by using the *second* derivative test? ....

....

....

6. The Warning on p169 is important. Although local maxima and minima occur at critical points, not every critical point will render a local maximum or minimum.

7. Remember that a local maximum or minimum can occur at the end of an interval, eg $f(x) = x^2$ on [0,1] has a local minimum at $x = 0$ and a local maximum at $x = 1$.

Checking the concepts

1. True or False?

   - The point $t = 2$ is a critical point of $f(t) = t^2 - 4t + 3$.   T/F

   - $f'(0)$ is not defined for $f(x) = \frac{1}{x}$ and therefore f has a critical point at $x = 0$. (Careful here; is $x = 0$ in the domain of f?)   T/F

   - (1, 0) is a critical point of $f(t) = \ln t - t$.   T/F

   - The function $f(t) = (1.2)^t$ has no critical points.   T/F

   - If f changes direction then f' changes sign.   T/F

2. If $f'(1) = 0$ and $f''(1) = 3$ then f has a local .... at $x = 1$.
   If $f'(2) = 0$ and $f''(2) = -0.5$ then f has a local .... at $x = 2$.

3. Does the simple function $f(x) = x$ have a local maximum on $[0, 1]$? ....

4. Sketch a function that has a critical point at $x = 0$ but the derivative does not change sign at the point.

5. True/False?

   - If $f' = 0$ at a point then f has a local max/min in that point.   T/F

- If we look for a local minimum of f on an interval then we only have to investigate critical points of f and endpoints of the interval.   T/F

- f will either have a local maximum or a local minimum at every critical point of f. T/F

Checking examples

- **Example 1:** The important idea in this problem is that (if f is defined on the interval) between critical points (in this example these are also the turning points), the function *cannot* change its behavior, it either increases or decreases.

- **Example 2:** This example makes use of the Second Derivative Test. You only need to check what the sign of the second derivative is at a critical point. Note that this test does not apply if the second derivative is zero at a critical point. See, for example, that for $f(x) = x^2$, we have $f(0) = 0$ and this is a local minimum but for $f(x) = x^3$, although we also have $f(0) = 0$, this is not a local minimum or maximum, it is called an inflection point, discussed in the next section.

- **Example 3:** This function has the shape of a third degree polynomial. In general third degree polynomials have one local maximum and one local minimum. There are exceptions to this rule. Can you think of such a function?

  ....

- **Example 4:** Notice that this is a *derivative* graph. Critical points occur where the graph cuts or touches the t-axis.

Take note

Where the derivative function cuts the t-axis from *above*, the function has a local maximum. When the derivative function cuts the t-axis from *below*, the function has a local minimum. Where the derivative function touches the t-axis from above or below, the function has neither a local maximum nor a local minimum but the concavity does change. Here the function has an inflection point, discussed in the next section.

Problems

1, 3, 5, 7, 11, 16, 22

4.2   INFLECTION POINTS

CAREFULLY READ THROUGH THIS SECTION IN THE TEXTBOOK

Key points

- Defining an inflection point of f.

- Locating inflection points.

Knowing the Book

1. A point at which the graph of a function changes .... is called an **inflection** point of f.

2. The concavity of f changes at an inflection point, so the .... of f'' changes there.

3. At an inflection point f'' is either .... or ....

4. We say that not every point where $f'' = 0$ is an inflection point. What example illustrates this?

   ....

Checking the concepts

1. Name a well-known periodic function for which the concavity changes from positive to negative at $x = 0$. ....

2. List the **number** of local maxima/minima and the **number** of inflection points for the following well-known functions:

function	no. local max/min	no. inflection points
the logarithmic function	....	....
a parabolic function	....	....
the sine function ($0 \leq x \leq 2\pi$)	....	....
the power function $y = x^9$	....	....
an exponential function	....	....
a quartic (max number)	....	....

3. True/False?

- If f has an inflection point and f'' exists then $f'' = 0$ at that point. T/F

- If $f'' = 0$ in a point then f has an inflection point there.   T/F

Checking examples

- **Example 1:** Notice that the fact that $f''(3) = 0$ alone does not ensure an inflection point. You have to see from the graph that there is an inflection point, or you can test points to the right and the left to confirm your suspicions of an inflection point.

- **Example 2:** Notice here that at a turning point the first derivative changes sign and at an inflection point the second derivative changes sign.

- **Example 3:** This curve is called the logistic curve and is discussed extensively in Section 5.6.

- **Example 4:** This curve is called the surge function and is discussed extensively in Section 5.7.

- **Example 5:** What would the graph look like if you make a new container which has the shape of two of these put on top of each other? How many inflection points are there then?

Take note

To find a turning point, look for points where $f' = 0$ or f' does not exist and check the sign of f' to the left and to the right of the point.

To find an inflection point, look for points where $f'' = 0$ or f'' does not exist and check the sign of f'' to the left and to the right of the point.

Problems

2, 10, 13, 15, 19, 21, 24, 26

4.3   GLOBAL MAXIMA AND MINIMA

CAREFULLY READ THROUGH THIS SECTION IN THE TEXTBOOK

Key points

- Defining a global minimum and a global maximum.

- Finding a global maximum and a global minimum:
  - on an interval including endpoints
  - on an interval excluding endpoints or on the entire real line.

Knowing the Book

1. What is optimization? ....

2. The global maximum of a continuous function f occurs at a .... maximum or at one of the ....

3. To find the global maximum and global minimum on an interval including endpoints:
   Compare values of the function at all the .... points in the interval and at the ....

4. To find the global maximum and minimum on an interval excluding endpoints:
   Find the values of the function at all critical points and ....
   ....

5. Is it possible for a function not to have a global maximum? ....
   ....

Checking the concepts

1. True or False?

   - Every continuous function on an interval that excludes endpoints has a global maximum.

   - The point $t = 2$ is a critical point of $f(t) = t^2 - 4t + 3$.   T/F

   - The global maximum of a function can occur at more than one point. T/F

   - The global minimum of $f(x) = x$ on [0,1] is 0.   T/F

- The function $f(t) = 2^t$ on (-1,1) has a global minimum but no global maximum.   T/F

- The earth has a "global maximum" on top of Mount Everest.   T/F

2. Match the function and the description:

   (a) The function has neither a global maximum nor a global minimum.

   (b) The function has a global maximum but no global minimum.

   (c) The local maximum and local minimum of the function are also the global maximum and global minimum, respectively, of the function.

   1. $y = \ln x$ for $x > 0$.

   3. $y = \sin t$ on $0 \leq t \leq 2\pi$.

   4. $y = -x^2 + 2x - 3$ on the real line.

3. The function f on [1,4], has one critical point, $x = 3$. We find that $f(1) = 4$, $f(3) = 6$ and $f(4) = 2$. Then the global maximum is .... and occurs at $x = $ ....
   The global minimum is .... and occurs at $x = $ ....

4. Does the function $f(x) = \ln x$ have a global maximum on $0 < x \leq 1$? Does it have a global minimum?

5. A function has no critical points on an interval [1,3]. Does it mean that the function does not have a global maximum or a global minimum?

6. Does the simple function $f(x) = x$ have a local maximum on $[0, 1]$? .... Does it have a global maximum? ....

Checking examples

- **Example 1:** The endpoints of the interval are included, so the important idea is to check function values at the critical points and at the endpoints.

- **Example 2:** The function describes a *rate* (or a derivative), so photosynthesis is fastest where the function has a global maximum. Note also the manipulation with the exponential function and the use of the second derivative test.

• **Example 3:** This example requires careful thinking. The technique used here is often used in economics and you will encounter it later again in Section 4.5. A hint for understanding this technique is to use a ruler and place it on the leftmost line joining the origin and a point on the graph. By swinging the ruler to the right (fixing it at the origin) the slope becomes less until the ruler becomes a tangent. By swinging the ruler left again the slope increases again. Minimum slope is where the ruler is a tangent.

Problems

3 - 6, 7, 12, 13, 16, 21

4.4   PROFIT, COST, AND REVENUE

CAREFULLY READ THROUGH THIS SECTION IN THE TEXTBOOK

Key points

• Maximizing Profit.

• Maximizing Revenue.

Knowing the Book

1. Maximum profit occurs when Marginal .... = Marginal ....

2. If $R' > C'$ then we could increase our profits by ....-creasing production.
   If $R' < C'$ then we could increase our profits by ....-creasing production.
   If $R' = C'$ could we increase our profits? ....

3. Revenue = ... × ....

4. If $q = f(p)$ is a linear function and and we substitute $f(p)$ for q in the revenue function $R = pq$, we obtain a revenue function as a function of .... This revenue function is then quadratic.

5. If the cost function is cubic, the marginal cost function is ....
   If the revenue function is linear, the marginal revenue function is ....

Checking the concepts

1. Maximum profit occurs at a point where the marginal cost and marginal revenue graphs have .... tangents.

   Maximum profit occurs at a point where the marginal cost graph and the marginal revenue graphs ....

2. If $MR = R'(10) = 2.5$ and $MC = C'(10) = 2.0$ then for maximum profit more items .... (should/should not) be produced.

3. If the demand function is $p = 80 - 3q$ then the revenue function $R = pq$ as a function of the quantity q is $R = $ .... and as a function of the price p is $R = $ ....

   Will these parabolic functions have a maximum values or a minimum values? ....

Checking examples

- **Example 1:** This is not an easy example.

  - Carefully revise the reasoning of how the graphs of the marginal revenue and cost curves are constructed and look at typical shapes. Hint: Use your ruler to follow the slope of the cost curve.

  - To draw the profit curve: Remember that profit = revenue - cost. So the profit is the difference between the two curves.

  - A key point in this problem is the reasoning why the profit is a minimum at q_1 and a maximum at q_2.

    Complete the summary:

    If $\pi' < 0$ then the profit is ....-creasing and

    if $\pi' > 0$ then the profit is ....-creasing.

    But $\pi' = MR - MC$

    So if $MR - MC < 0$ (or $MR < MC$, the MR graph lies below the MC graph) then the profit is ....-creasing and

    if $MR - MC > 0$ (or $MR > MC$, the MR graph lies above the MC graph) then the profit is ....-creasing.

    To the right of q_1 $MC < MR$, so profit increases if one moves away from q_1.

    To the right of q_2 $MC > MR$, so profit increases if one moves towards q_2.

- **Example 2:** Another way of approaching this problem is to find the formula for the profit function $\pi(q) = R(q) - C(q) = -0.003q^2 + 3.9q - 300$. This is a concave down parabola. The local maximum is at $q = 650$ and the global minimum is either at $q = 0$ or at $q = 1000$. Checking function values reveals that the global minimum is at $q = 0$.

- **Example 3:** Notice that the revenue function is a concave down parabola for which the local maximum is also a global maximum.

- **Example 4:** In the previous example a revenue function was drawn up as a function of the quantity because of the question "What quantity?".
  In this example the question is "What price?" so a revenue function as a function of price is drawn up.

Problems

2, 3, 5, 6, 9, 13,

4.5   AVERAGE COST

CAREFULLY READ THROUGH THIS SECTION IN THE TEXTBOOK

Key points

- Explaining what average cost is.

- Visualizing average cost.

- Minimizing average cost.

- The relationship between average cost and marginal cost.

Knowing the Book

1. Average cost is cost divided by ....
   As a formula:

   ....

2. The average cost to produce q items can be visualized as the .... of the line from the origin to a point $(q, ...)$ on the cost curve.

3. Where *marginal cost* equals *average cost* is where average cost could be a
....

4. Marginal cost and average cost:

   • If marginal cost is greater than average cost, then by increasing production average cost is ....-creased.

   • If marginal cost is less than average cost, then by increasing production average cost is ....-creased.

   • What happens when marginal cost equals average cost? ....
....
Draw a graph to illustrate these statements.

Checking the concepts

1. The units of average cost are ....
The units of marginal cost are ....

2. If $a(100) = 3.15$ and $C'(100) = 4.17$ then average cost increases / decreases.

3. (a) If $C(50) = 300$ then $a(q) =$ ....

   (b) If $C'(50) = 4.5$ should the 51st item be manufactured for lowering the average cost? ....

4. If $C(q) = q^3 - 6q^2 + 12q + 4$, C in thousands of dollars, and if $C'(q) = 3q^2 - 12q + 12$ then

   • $a(10) =$ ....

   • $C'(10) =$ ....

Will the average cost increase as production increases? ....

Checking examples

- **Example 1:** A simple example to illustrate the calculation of average cost.

- **Example 2:** The conclusion here is that average cost for producing 100 items is greater than marginal cost for this example. Why is this good, what will happen to the average cost if production is increased? ....
  ....

- **Example 3:** Similar to Example 3 of Section 4.3. A ruler might be handy here again.

- **Example 4:** This example uses numbers to illustrate the rules in the block just below (top of next page). Make sure you follow the reasoning.

- **Example 6:** This example covers all the concepts dealt with in this section. Work through it with care.

Problems

1, 2, 5, 7, 9, 10

4.6   ELASTICITY OF DEMAND

CAREFULLY READ THROUGH THIS SECTION IN THE TEXTBOOK

Key points

- Defining elasticity of demand.

- The relationship between elasticity of demand and revenue.

Knowing the Book

1. The elasticity of demand E is the .... of fractional ....
   .... to fractional ....

2. The formula for calculating E is

   ....

   NB: Changing the price of an item by 1% causes a change of ....% in the quantity of goods sold.

3. When is the demand elastic? ....
   When is the demand inelastic? ....

4. Say the price of an item is **raised**.
   If demand is inelastic ($E < 1$), revenue will ....-crease.
   If demand is elastic ($E > 1$), revenue will ....-crease. (In this case it will be better to .... the price.)

5. If $E < 1$ revenue is increased by .... the price.
   If $E > 1$ revenue is increased by .... the price.

6. NB: If $E = 1$ we find the critical points of the revenue function and so have (possibly) located where the revenue function is maximised.

Checking the concepts

1. Can the elasticity of demand, E, be negative? ....

2. What are the units of E? ....

3. What does it mean if $E = 2$? ....
   ....

4. Say $p = \$20$ and $q = 50$.

   - Let $E = 4$. If p increases by 1% to $20.20 then the demand will decrease by ....
     to ....

   - Let $E = 4$. If p decreases by 1% to $19.80 then the demand will ....
     ....

   - Let $E = 0.5$. If p increases by 1% to $20.20 then the demand will ....
     ....

   - Let $E = 0.5$. If p decreases by 1% to $19.80 then the demand will ....
     ....

5. Say, once again, that $p = \$20$ and $q = 50$.

   - The revenue $R = $ ....

   - If $E = 4$ and p increases to $20.20 the demand decreases to 48 (decrease of 4% of 50). The revenue ......-creases to ....
     The price should be .... to increase the revenue.

- If $E = 0.5$ and then if p increases to \$20.20 the demand decreases to 49.75. The revenue ......-creases to ....

    The price should be .... to increase the revenue.

Checking examples

- **Example 1:** This example does not make use of the derivative $\frac{dq}{dp}$ because it cannot be calculated from the information given. The approximation $\frac{\Delta q/q}{\Delta p/p}$ is used instead.

- **Example 2:** In this case the derivative is given and the true formula for E can be used.

- **Example 3:** A bit of a theoretical example that makes use of the product rule. It explain where the formula for elasticity of demand comes from.

Problems

1, 2, 4, 5, 6, 10, 13, 17

4.7 LOGISTIC GROWTH

CAREFULLY READ THROUGH THIS SECTION IN THE TEXTBOOK

Key points

- Modeling the US population as an example of logistic growth.

- Defining the logistic function.

- Sales Predictions.

- Dose-Response Curves.

Knowing the Book

1. Population with an upper bound can be modeled with a *logistic* or .... model.

2. A **logistic function** is a function in the form

    ....

    where .... are .... constants.

3. The value L is called the ....
   of the environment.

4. The parameter k effects the .... of the curve. (The
   bigger the value of k the .... the curve.)

5. The point of ....
   is where the concavity changes, and it occurs where $P =$ ....

6. The logistic curve is approximately exponential for small values of t, with
   the growth rate roughly equal to k. The curve can then be approximated by
   $P = P_0 e^{\cdots\cdots t}$.

7. The independent variable for a dose-response curve is the ....,
   (not time).
   On the y-axis, the ....
   is given, with a maximum value of ....

8. What are the advantages of administering a drug with a not-so-steep dose-
   response curve to administering a drug with a steep dose-response curve.
   ....
   ....

Checking the concepts

1. Match the function and the description:

(a)	$P = \frac{100}{1+10e^{-0.03t}}$		(1)	initial value of 5
(b)	$P = \frac{50}{1+9e^{-0.04t}}$		(2)	not a logistic function
(c)	$P = \frac{80}{1+10e^{0.03t}}$		(3)	a growth rate of 3% for small t
(d)	$P = \frac{60}{1+4e^{-0.04t}}$		(4)	a carrying capacity of 70
(e)	$P = \frac{70}{1+19e^{-0.05t}}$		(5)	changes concavity when $P = 30$

2. If a logistic curve changes concavity at the point $(8, 320)$, the carrying capacity
   is $L =$ ....

3. Is it possible for a logistic function to be 0 when $t = 0$?
   ....

4. Mark the following on the dose-response curve below:

   • A dose of 5 mg will cause a response of 30%

- A dose of 10 mg will cause a response of 80%

- A dose of 15 mg will cause a response of 98%

intensity of response

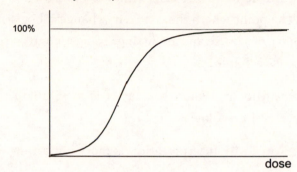

5. If, on the dose-response curve above, a safe and effective response is between 40% and 70%, then (estimated from the graph) a safe dose to administer is between .... and ....

6. For the two dose-response curves below, what is the minimum dose for each curve that will trigger the minimum response (approx)?
   (a) ....
   (b) ....

intensity of response

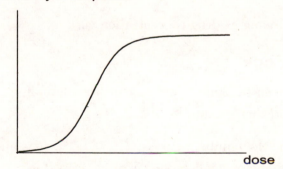

Checking examples

- **Example 1:** On the graphs in this example it seems as if $P = 0$ if $t = 0$. This is not the case. The value of P (for $L = 4$, for example) when $t = 0$ is .... Notice that k has no influence on the initial value of P.

- **Example 2:** There seems to have been a surge in the US population since 1940. It is interesting to sketch the shape of the US population curve from 1790 up to 1990.

- **Example 3:** Notice that with any logistic curve the rate of change is maximum at the point where the concavity changes. Logistic regression is mentioned here which of course means fitting a logistic curve to a set of data. For this one needs technology.

- **Example 4:** Would you say it is inevitable that this exponential curve will develop into a logistic curve?

- **Example 5:** Interpretation is more important here than calculations.

Problems

1, 3, 4, 5, 8, 11, 13

4.8   THE SURGE FUNCTION AND DRUG CONCENTRATION

CAREFULLY READ THROUGH THIS SECTION IN THE TEXTBOOK

Key points

- Introducing the family of functions: $y = ate^{-bt}$ and discussing the effect of the parameters a and b.

- Looking at drug concentration curves.

Knowing the Book

1. When smoking, a person's nicotine level first ....
   and then ....

2. For the family of curves, $f(t) = te^{-bt}$: (b is always ....-tive)
   As b gets smaller, the curve increases for a .... time
   and to a .... value.

3. The point where the function $y = te^{-bt}$ changes from ....-creasing to
   ....-creasing is $t =$..............

4. The *drug concentration curve* plots the concentration of ....
   .... against ....

5. For a drug concentration curve the highest concentration is called the ....
concentration.

6. What is the drug concentration in the blood necessary to achieve a pharma-cological response called? ....
The time taken to reach the *minimum effective concentration* is referred to as the ....
The time between the onset and the termination of pharmacological response is the .... of .... (See Figure 4.80)

Checking the concepts

1. For $y = 3te^{-0.25t}$,

   - $b = $ .... and $a = $ ....
   - the function changes from increasing to decreasing at $t = $ ....
   - the function reaches a maximum value of ....

2. For a drug concentration curve $C = ate^{-bt}$, t is measured in .... (for example) in C in .... (for example).

3. Which of the two curves:

   (a) $C = te^{-t}$ or

   (b) $C = te^{-2t}$

   reaches peak concentration first? ....
   reaches the highest peak concentration? ....

4. Which of the two curves:

   (a) $C = te^{-t}$ or

   (b) $C = 2te^{-t}$

   has the longest duration of effectiveness? ....

5. For the drug concentration curve below, t in hours, peak concentration occurs after 1 hour. An estimate for

   (a) the onset of pharmacological response is ....

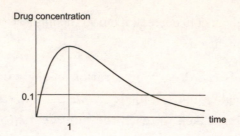

(b) the termination of pharmacological response is ....

(c) the duration of effectiveness is ....

Checking examples

- **Example 1:** When finding the derivative and setting it equal to zero remember that an exponential function can never be zero.

- **Example 2:** Would it do good or harm to administer the paracetamol with propantheline to the group of slow absorbers? ....

- **Example 4:** What would happen if the dosage of 150mg is lowered? ....
  ....

Problems

2, 4, 5, 7, 10

REVIEW OF CHAPTER 4

1. True/False?

   (a) A function has a local maximum or a local minimum at every critical point.   T/F.

   (b) A function can have more than one local minimum.   T/F.

   (c) If π is the profit function and the derivative π' changes from negative to positive, then the profit is at a local minimum at that point.   T/F.

   (d) If the average cost of producing 250 items is $45 per item and the marginal cost is $48, an increase in production would lower the average cost. T/F.

2. If f' changes sign at p then f has a ....
.... at p.

3. The concavity of a function changes from negative to positive. Sketch the rate of change of the function near the inflection point.

4. Once you've located all the local maxima on $a \leq x \leq b$, how would you then go about locating the global maximum of the function on the interval? ....
....

5. Say f has one inflection point and it occurs at $t = 3$. If, in addition, we know that $f''(1) = 2$, then f changes from concave .... to concave
....

6. If f' is a positive but decreasing function, then f is ....-creasing at a ....-creasing rate.

7. For each of the following:
   - Draw a graph that describes the situation, marking the axes appropriately.
   - Use symbols to describe the situation, substituting number values for parameters where possible.

   • Your energy level peaks 1 hour after taking an energy supplement.

• An epidemic spreads through a city with a population of 400 000.

....

Review Problems

6, 9 - 12, 13, 17, 20, 29, 32

CHAPTER 5

ACCUMULATED CHANGE: THE

DEFINITE INTEGRAL

5.1 ACCUMULATED CHANGE

CAREFULLY READ THROUGH THIS SECTION IN THE TEXTBOOK

Key points

- Measuring distance traveled:

  - for constant velocity
  - for changing velocity

- Approximating total change from rate of change.

Knowing the Book

1. The question answered in this chapter is: *If we know the ....* *change, how can we determine the .... change?*

2. Another term for *accumulated change* is .... change.

3. The rate of change of distance is ....

4. Graphically: It appears that the distance traveled is equal to the area under the .... graph.

5. If velocity is changing continually: We can approximate distance traveled by assuming that the velocity is .... over small intervals.

6. By *halving* the interval of measurement, we .... the difference between the upper and the lower estimates.

Checking the concepts

1. If a parachutist's velocity increases from 5 ft/sec to 9 ft/sec in 1 second, then during that 1-second period he travels at *least* .... and at *most*................ (Remember to include units).

2. If the parachutist's velocity increased from 5 ft/sec to 6.5 ft/sec in half a second and increased further to 9 ft/sec in another half a second, then the parachutist traveled at least .... during the 1-second period and at most ....

3. Give a graphical presentation of each of the two cases above:

4. If a car's velocity decreased from 38 ft/sec to 22 ft/sec in 2 seconds, then the car traveled at least .... in the two second period and at most ....

5. Given the following two statements:

   (a) At the start of 1990 Sarah was 39 inches tall and at the start of 1992 she was 43.5 inches tall.

   (b) At the start of 1990 Sarah grew at a rate of 2 inches per year and at the start of 1992 she grew at a rate of 3 inches per year.

   Say you want to calculate the *change* in Sarah's height between 1990 and 1992: In which case can the change be calculated accurately? .... What is the change? ....
   In the other case change can only be estimated, namely that it lies between .... and ....

Checking examples

- **Example 2:** In this example we assume that the population grew at a *constant* rate of 5000 people/year for 3 years and then the rate suddenly decreased to 3000 people/year and stayed that way for 4 years. This situation is unlikely but illustrates the relationship between a rate of change and total change.

- **Example 3:** In this table only a few function values are given. The graph is a set of points and so we can only *estimate* the total sales.

    - Also see Problem 6 at the end of the section.

    - Note that the first rectangle of the underestimate has a height of zero and so does not contribute to the area.

– In this example the notion is introduced that the best *single* estimate of total change is the average between the overestimate and the underestimate.

Problems

2, 6, 7, 9, 11

5.2   THE DEFINITE INTEGRAL

CAREFULLY READ THROUGH THIS SECTION IN THE TEXTBOOK

Key points

- Introducing n and Δt.

- Improving the approximation by decreasing Δt.

- Introducing left- and right-hand sums.

- Taking the limit to obtain the definite integral.

- Estimating a definite integral from a table or a graph.

Knowing the Book

1. We use the notation Δt for the .... of a subinterval and n for the .... of subintervals.

2. The number of subintervals also represent the number of .... in the graphical presentation.

3. If we take measurements in the time interval $a \leq t \leq b$ at equally spaced times $t_0, t_1, t_2, ..., t_n$ with time $t_0 = a$ and $t_n = b$ then the length of the time interval between two consecutive measurements is

$$\Delta t = \text{...................}$$

4. The *left-hand sum* for **estimating** total change between a and b is given by:

....

The *right-hand sum* for **estimating** total change between a and b is given by:

....

5. If f is an increasing function, the left-hand sum will be an ....-estimate of the total change.

   If f is a decreasing function, the left-hand sum will be an ....-estimate of the total change.

6. As n gets larger, the sum of the rectangles approaches the .... under the curve.

7. If n is large enough, both the left-hand sum and the right-hand sum are accurate estimates of the ....

8. The **definite integral** of f from a to b, written

   ....,

   is the limit of the left-hand **or** right-hand sums with n subdivisions as n gets

   ....

   Each of the left- and right-hand sums is called a .... sum, f is called the

   ...., and a and b are called the ....

   ....

Checking the concepts

1. For a given time interval, if Δt *decreases* then n ....-creases and the estimate for total change should .... (improve / deteriorate)

2. For the interval $2 \le t \le 4$ and $\Delta t = 0.5$ then $n =$ .... and $t_0 =$ ...., $t_1 =$ ...., $t_2 =$ ...., $t_3 =$ .... and $t_4 =$ ....

3. For the interval $3 \le t \le 5$ and $n = 10$ then $\Delta t =$ .... and $t_0 =$ ...., $t_1 =$ ...., $t_2 =$ ...., $t_3 =$ ...., $t_4 =$ .... $t_9 =$ .... and $t_{10} =$ ....

4. For the interval $2 \le t \le 4$ and given $\Delta t = 0.5$ and $f(t) = t^2 + 3$ then $f(t_0) =$ ...., $f(t_1) =$ ...., $f(t_2) =$ ...., $f(t_3) =$ .... and $f(t_4) =$ ....

5. Sketch the decreasing function $f(t) = (0.8)^t$ over the interval $0 \le t \le 10$. Take $\Delta t = 2$ and sketch both the left-hand sum and the right-hand sum.

Which is the upper estimate and which is the lower estimate?

....

....

6. True or false?

(a) n indicates the number of rectangles in the left-(or right) hand sum. T/F

(b) If f increases then the left-hand sum is a lower estimate.   T/F

(c) It does not really matter whether you work with the left- or right-hand sums because both will approach the value of the definite integral as n gets large.   T/F

(d) The value of the definite integral is always between the values of the left-hand sum and the right-hand sum.   T/F

(e) Total change can be seen as the area under the graph of the rate of change function (for a positive rate of change).   T/F

Take note

When sketching a left-hand sum you work from left to right across a sub-interval and when you draw a right-hand sum you work from right to left across the sub-interval.

Checking examples

- **Example 1:** Notice that the function describes a *rate of change* and the units are 'million of bacteria per hour'. $f(3) = 3.9$ tells you that the bacteria population grows at a rate of 3.9 million of bacteria per hour.

- **Example 2:** This example illustrates the use of a calculator that can do integration.

- **Example 3:** The best we can do is to work out the left-hand sum and the right-hand sum and take the average as was suggested in the previous section.

- **Example 4:**
Function values can only be estimated in this example because it is read from a graph. For accurate function values we need either a formula or a table.
Note that again we have a decreasing function, and so the right-hand sum is a lower estimate.

Problems

1, 2, 3, 6, 9, 11

5.3   THE DEFINITE INTEGRAL AS AREA

CAREFULLY READ THROUGH THIS SECTION IN THE TEXTBOOK

Key points

- Using the definite integral to calculate an area:

  - when $f(x)$ is positive.
  - when $f(x)$ is not positive.

- Calculating the area between two curves.

Knowing the Book

1. If $f(x)$ is positive (the graph lies above the x-axis) and $a \leq b$ then the area under the graph of f between a and b is given by:

   ....

2. If a graph lies below the x-axis, the definite integral is the .... of the area.

3. When $f(x)$ is positive for some x values and negative for others then:
   $\int_a^b f(x)dx$ is the sum of the areas above the x-axis, counted ....
   and the areas below the x-axis, counted ....

4. Area between graphs: If $g(x) \leq f(x)$ for $a \leq x \leq b$ then the area between the graphs of $f(x)$ and $g(x)$ between a and b is given by:

   ....

Checking the concepts

1. The graph of $f(t) = (1.3)^t$ lies .... (above / below) the x-axis. Therefore the area enclosed between the graph of f and the x-axis between $x = 0$ and $x = 2$ is given by ....

2. Given that $\int_0^2 (2x - 4)dx = -4$.

   (a) Sketch the function $f(x) = 2x - 4$ on $[0,4]$ and explain the negative answer above from the graph.

   (b) Given that $\int_2^4 (2x - 4)dx = 4$ and $\int_0^4 (2x - 4)dx = 0$. What is the graphical interpretation of these integral values? (Use the sketch above.)

   (c) For the function $f(x) = 2x - 4$ the *area* enclosed between the graph of f and the x-axis between $x = 0$ and $x = 4$ is ....

3. If one has to find the area enclosed between the function $f(x) = x^2 - 9$ and the x-axis between $x = 0$ and $x = 6$ (a sketch might be useful), then you first integrate f between .... and .... and then integrate f between .... and .... and then ....
   ....

4. True/False?

   • An integral can be negative, an area never.   T/F

   • The sum of two areas can be zero.   T/F

   • The sum of two integrals can be negative.   T/F

   • If you're asked to calculate an area it is wise to determine the roots of the function first.   T/F

   • The area enclosed between two curves can be negative.   T/F

Take note

When calculating the area enclosed between two curves, one should remember the "top minus bottom" rule:

The area between two curves is the integral of the top function minus the bottom function.

What will happen if the curves f and g intercept and change roles over the interval? The answer to this question is that you will have to integrate $f - g$ (if $f \geq g$ at first) up to where they intercept and from there you integrate $g - f$.

Example: How would you go about calculating the area between $f(x) = x$ and $f(x) = x^2$, between $x = 0$ and $x = 2$? (A sketch should be useful.)

The functions intercept at $x = 1$. The required area is given by:

$$\int_0^1 (x - x^2)dx + \int_1^2 (x^2 - x)dx$$

Checking examples

- **Example 1:** From the sketch it is clear that the value of 2 may be a bit low. The shaded area above this line is not quite as big as the white areas below the line. A value of 2.25 might be better; the estimate of the area then is 2.25×3 which is closer to the true value of 6.967.

- **Example 2:** Note the difference between the terms *the integral* and *the area*.

- **Example 3:** This example illustrates that the value of an integral can be the difference between two areas.

- **Example 5:** In this example the word *area* is not mentioned. We only look at integrals which can be positive, negative or zero.

- **Example 6:** The phrase "estimate the area.." indicates that whatever the answer to the problem may be, it has to be a positive number because area can only be positive. Also note the importance of finding the root of the function whenever an area has to calculated over an interval that includes a root of the function.

Problems

5 - 8, 17, 21, 23 - 26

5.4 INTERPRETATIONS OF THE DEFINITE INTEGRAL

CAREFULLY READ THROUGH THIS SECTION IN THE TEXTBOOK

Key points

- Notation and units for the definite integral.

- Interpreting a definite integral.

- Discussing bioavailability of drugs.

Knowing the Book

1. The units of measurement for $\int_a^b f(x)dx$ is the product of the units of ....
   and the units of ....

2. If $f(t)$ is a rate of change of a quantity, then $\int_a^b f(t)dt$ represents ....
   ....

3. What are typical units of bioavailability? ....

Checking the concepts

1. Given a list of rates of changes (f) and their units, list the units of the integral:

Description of f	Units of f	Units of x	Units of $\int_a^b f(x)dx$
child's growth rate	inches per year	years	....
acceleration	feet per sec per sec	sec	....
marg building cost	dollars per square feet	sq feet	....
factory productivity	items per person	people	....

2. Match the following:

The rate of change graph ...	**The function graph is..**
1. is positive but decreasing.	a. has a turning point.
2. cuts the x-axis.	b. decreasing and concave down.
3. dips further below the x-axis.	c. a straight line.
4. is constant.	d. increasing at a decreasing rate.

3. True/False?

   If, over an interval,

   - the rate of change graph lies below the x-axis then the quantity decreases over the interval.   T/F

   - the rate of change is negative then the total change is negative.   T/F

   - the rate of change graph is a horizontal line then the total change is zero.   T/F

Take note

Rate of change is normally given in the form $P'(t)$ or $\frac{dy}{dt}$, but remember that the derivative is also a function. Therefore the rate of change is often given as a function $f(t)$.

Checking examples

- **Example 1:** An important concept in this example is that the population size at time $t = 2$ is the population size at $t = 0$ plus the change over this period.

- **Example 3:** This is an important example. The function tells us what the velocity is at which the man travels, and the integral tells us what the distance is (how far from or near to the house).

  - Tempting here is to say that the car turns around when the graph has a turning point. Not so, the car turns around when the graph cuts the x-axis. During the first two hours the man accelerates and then slows down. He then accelerates in the opposite direction and then slows down.

  - See if you can answer the following questions on the velocity:
    * When was the man driving most rapidly towards the house? ....
    * When was the man driving most rapidly away from the house? ......
    * When did the man turn his car around? ....

- **Example 4:** The important thing here is to look at *areas* and not at function values because *rate of change* curves are given.

- **Example 5:** Bioavailability is given by the area under the drug concentration curve (the surge function).

Problems

2, 3, 5, 7, 11, 20

5.5   THE FUNDAMENTAL THEOREM OF CALCULUS

CAREFULLY READ THROUGH THIS SECTION IN THE TEXTBOOK

Key point

Using the Fundamental Theorem of Calculus to find the total change in F between a and b.

Knowing the Book

1. The **Fundamental Theorem of Calculus** states that:

   If F' is .... for $a \leq t \leq b$ then

   $$\int_a^b F'(t)dt = ...$$

   In words: The definite integral of a derivative gives ....
   ....

2. The cost to increase production from a units to b units is given by

   ....

3. The total variable cost to produce b units is given by

   ....

   The total cost to produce b units is given by

   ....

Checking the concepts

1. Complete the following:

   (a) $\int_2^4 F'(t)dt = ... - ....$

   (b) $\int_3^5 4xdx = ... = ...$ $(\frac{d}{dx}(2x^2) = 4x)$

   (c) $F(10) - .... = \int_2^{10} ....dt$

   (d) If $f(t) = \frac{d}{dt}(5t^2)$ then $\int_0^3 f(t)dt = ... = ...$

   (e) $.............. = F(5) + \int_5^7 f(t)dt$ if $F' = f$.

2. While the rate of change is positive, the total change is ....-creasing.

3. Say what each of the following means:

   $C'(40) = 5$ ....

   $\int_{40}^{50} C'(q)dq = 55$ ....

$$\int_0^{50} C'(q)dq = 105 \quad \text{..}$$

$$C(0) + \int_0^{50} C'(q)dq = 150 \quad \text{...}$$

Checking examples

- **Example 1:** It is important here to look at where F' lies above the t-axis and where it lies below. Because the curve lies below the t-axis during the first three months the total change (the integral) is negative and the investment decreased. Make sure that you understand this important example.

- **Example 2:** Remember what the graph of the usual cost function looks like? Does it make sense that the marginal cost curve has this shape?

Problems

1, 3, 4, 12

REVIEW OF CHAPTER 5

1. For the interval $3 \leq t \leq 4$ and $\Delta t = 0.25$ then $n = $ .... and
$t_0 = $ ...., $t_1 = $ ...., $t_2 = $ ...., $t_3 = $ ...., $\cdots t_n = $ ....

2. The integral $\int_3^4 f(t)dt$ can be approximated by a left-hand sum as follows (taking $\Delta t = 0.25$):

....

3. For the function f sketched below on $2 \leq t \leq 4$, say what the following expressions each present:

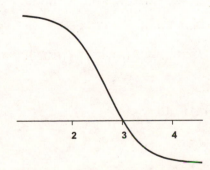

- $0.5(f(2.5) + f(3) + f(3.5) + f(4))$

....

- $\int_2^3 f(x)dx - \int_3^4 f(x)dx$

....

- $0.25(f(2) + f(2.25) + f(2.5) + f(2.75))$

....

4. The definite integral of a rate of change gives ....

5. If $\int_2^4 f(x)dx = 0$, but f is not the zero-function, how can you explain the 0?

....

6. The area included between the graphs of $f(x) = x^3$ and $g(x) = x$ between $x = 0$ and $x = 2$ is given by the expression

....

7. If $C'(q)$ represents marginal cost, then

$$\int_{100}^{250} C'(t)dt$$

represents ....

and the units of the integral expression are ....

8. How would you go about calculating $\int_0^{10} f(t)dt$ if

- values of $f(0), f(1), f(2), \cdots, f(9), f(10)$ are given in a table.

....

- a formula for $f(t)$ is known (two or three methods).

  – ....

  – ....

  – ....

Review Problems

1, 7, 10, 11, 14, 16, 19

FOCUS ON THEORY

THEOREMS ABOUT DEFINITE INTEGRALS

CAREFULLY READ THROUGH THIS SECTION IN THE TEXTBOOK

Key points

- Using the Second Fundamental Theorem of Calculus.

- Looking at properties of the Definite Integral.

Knowing the Book

1. The Second Theorem of Calculus: If f is a continuous function on an interval, and if a is any number on that interval, then the function G defined by

$$G(x) = \int_a^x f(t)dt$$

has a derivative ......, that is $G'(x) = $ ....

2. Write in symbols:

- The integral from a to b plus the integral from b to c is the integral from a to c:

   ....

- The integral of the sum of two functions is the sum of the integrals of the two functions.

   ....

- The integral of a constant times a function is the constant times the integral of the function.

   ....

Checking the concepts

1. Why is $G(x + h) - G(x) \simeq f(x)h$ if $G'(x) = f(x)$?

2. Sketch $f(x) = x$ for $0 \leq x \leq 3$. Find $G(x) = \int_0^x f(x)dx$ for $x = 0$, $x = 1$, $x = 2$ and $x = 3$ geometrically.

   Plot the points that you have calculated and draw a curve through the points. Does it seem as if $G'(x) = f(x)$?

3. True/False?

   - $\int_0^5 xe^x dx = \int_0^5 xdx \times \int_0^5 e^x dx$ T/F

   - $\int_0^5 0.1xe^x dx = 0.1 \int_0^5 xe^x dx$ T/F

   - $\int_0^5 (x - e^x)dx = \int_0^5 xdx - \int_0^5 e^x dx$ T/F

   - $\dfrac{1}{\int_0^5 e^x dx} = \int_0^5 e^{-x} dx$ T/F

   - $\int_0^5 e^x dx - \int_0^2 e^x dx = \int_2^5 e^x dx$ T/F

Problems

1, 2, 3, 5, 9-11

CHAPTER 6

USING THE INTEGRAL

6.1   AVERAGE VALUE

CAREFULLY READ THROUGH THIS SECTION IN THE TEXTBOOK

Key point

- Calculating the average value of a function by using a definite integral.

Knowing the Book

1. The average value of f from a to b is given by:

    ....

2. If we interpret the integral as the area under the graph of f: The average value of f is the .... of the rectangle whose width is .... and whose area is the same as the area between the graph of f and the x-axis.

    or
    $$\int_a^b f(x)dx = (\text{Average value of } f) \times \text{..}$$

Checking the concepts

1. Match the function and its average value over the interval $0 \leq x \leq 4$:

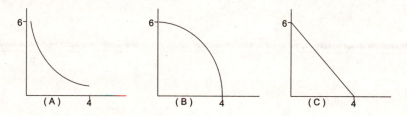

    (1)   the average value is 3   ....
    (2)   the average value is less than 3   ....
    (3)   the average value is more than 3   ....

2. Sketch $f(x) = t^2$ for $0 \leq x \leq 1$. Then sketch a rectangle with the same area as the area under f. Estimate the height of the rectangle.

3. Can the average value of f from a to b be negative? ....

4. True/False?

   - If a function cuts the x-axis on an interval it is possible that the average value of the function over the interval is zero.   T/F

   - When calculating an average value it is important to calculate the roots of the function.   T/F

   - The average value of $f(x) = x^2$ on the interval $0 \le t \le 1$ is $\frac{1}{2}$.   T/F

5. Easy averages:

   - The average value of the line $y = x$ on $[0, 4]$ is ....

   - The average value of $y = \sin t$ over one period is ....
     The average value of $y = 20 + 5\sin t$ over one period is ....

Checking examples

- **Example 1:** Note how the units of the average value of f are shown to be the same as the units of f.

- **Examples 2 & 3**: These examples show two ways of calculating an average value. The first one makes use of a formula and the second one makes use of a graph.

Problems

2, 5, 9, 12, 18

6.2   CONSUMER AND PRODUCER SURPLUS

CAREFULLY READ THROUGH THIS SECTION IN THE TEXTBOOK

Key points

- Defining producer surplus and consumer surplus.

- Discussing the effect of wage and price controls.

Knowing the Book

1. As the price increases the quantity supplied goes .... and the demand for the product goes ....

2. What does the **consumer surplus** measure? ....
   ....

3. Explain what the consumer surplus is: ....
   ....
   ....

4. What does the **producer surplus** measure? ....
   ....

5. Explain what the producer surplus is: ....
   ....
   ....

6. The **consumer surplus** is given by the area between the ....
   curve and the line ....

7. The **producer surplus** is given by the area between the ....
   curve and the line ....

8. If the price is artificially high the consumer surplus ....-creases.

9. Total gains from the trade is given by
   .... + ...
   If the price is kept artificially high the total gains from the trade ....-creases.

Checking the concepts

1. What does it mean if $D(200) = 50$ for a particular demand curve? ....
   ....
   What does it mean if $S(200) = 20$ for a particular supply curve? ....
   ....
   At the equilibrium point: Will p^* be more or less than \$50? ....
   Will q^* be less or more than 200? ....

2. For the demand curve $D(q) = 100e^{-0.004q}$, and equilibrium quantity $q = 1000$

- the equilibrium price is ....

- the consumer surplus is given by ....

3. Say a producer surplus works out at $2300. What does this tell you?

   ....

   ....

4. Say a consumer surplus works out at $1340. What does this tell you?

   ....

   ....

Take note

Remember that **c**onsumers **d**emand and **p**roducers **s**upply. The letters in the first half of the alphabet (c & d) belong together and so do the letters in the second half (p & s). Also, the **d**emand curve **d**ecreases and the **s**upply curve **s**oars.

Checking examples

- **Examples 1 & 2:** There are only two examples in this section. Both of these are comprehensive. It will be well worth your while to spend a fair amount of time on these examples. First make sure, however, that you are comfortable with the terminology and the concepts.

Problems

1, 2, 3, 7, 13

6.3   PRESENT AND FUTURE VALUE

CAREFULLY READ THROUGH THIS SECTION IN THE TEXTBOOK

Key points

- Calculating the present value of a continuous stream of payments.

- Calculating the future value of a continuous stream of payments.

Knowing the Book

1. The income stream is a .... at which payments are made and an example of the units of an income stream are ....

2. The future value represents the total amount of money that you would have at a future date if you ....
....

3. The present value represents the amount of money you would have to deposit today in order to match ....
....

4. If you deposit B dollars t years from now at an interest rate of r, the present value is ....

5. If deposits are being made at a rate of $S(t)$ then the total present value of the amount paid in over a period of M years is
....

6. M years into the future at a rate of r

Future value = Present value × ....

Checking the concepts

1. A constant stream of income of $S(t) = 1200$ dollars per year means that per six months the income is .... and per month the income is ....

2. At present your investment is valued at \$50,000. What will its value be 5 years from now at a continuous interest rate of 6%? ....

3. You will inherit \$10 000 in one year. What is the present value at a 6% interest rate? ....

4. Income streams are not always constant, for example $S(t) = 1200t$ means that initially the income stream is .... dollars per year and after 1 year it is .... dollars per year.

5. An income stream of $S(t) = 100 + 10t$ means that 1 year from now the income flows in at a rate of .... dollars per year and two years from now the money flows in at a rate of .... dollars per year. The present value of money flowing in during the first year earning 6% interest is given by the integral

....

and the present value of money flowing in during the second year earning 6%

interest is given by

....

6. For the income stream above, the future value of money flowing in over the first two years and earning 6% interest is given by

....

Take note

An income stream is a *rate of change* and income is the *integral* of the income stream.

Checking examples

- **Examples 1:** This example is a direct application of the formulas of present and future value

- **Examples 2:** This example illustrates the difference between depositing a lump sum and a continuous deposit stream. It is hardly possible for someone to deposit continuously, therefore the approximation of $480 per month at the end of the example.

Problems

1, 2, 4, 8, 10

6.4   RELATIVE GROWTH RATES

CAREFULLY READ THROUGH THIS SECTION IN THE TEXTBOOK

Key points

- Pointing out the difference between:
  - Absolute rate of change.
  - Relative rate of change (relative growth rate).

- Computing changes from the relative growth rate.

Knowing the Book

1. For a function P of t the *absolute* rate of change of P with respect to t is given by:

   ....

2. For a function P of t the *relative* rate of change of P with respect to t is given by:

   ....

3. Relative growth rate is measured in .... per year.

4. Given the relative growth rate, we cannot find the change in population size, but we can find the .... change.

5. The integral of the relative growth rate gives the total change in ....
   In symbols:

   $$\int_a^b \frac{P'(t)}{P(t)} = ...$$

Checking the concepts

1. For $P = 100e^{0.1t}$

   - The absolute growth rate is given by:

     $\frac{dP}{dt} =$ ....

   - The relative growth rate is given by:

     $\frac{P'(t)}{P(t)} =$ .... (a percentage)

2. For a population $P = 100e^{0.1t}$, t in years:

   - The integral of the relative growth rate over this three year period is given by:

     ....

     which means that $\ln P(3) - \ln P(0) =$ ....

     so that $\frac{P(3)}{P(0)} =$ ....
     The population has increased by a percentage of .... over the first three years.

3. $\int_0^5 \dfrac{P'(t)}{P(t)}\,dt =$ .... $-$ ....

4. If $\dfrac{P(3)}{P(0)} = 3$, has the population increased by 200% or 300% over this three year period? ....

5. $\ln P(50) - \ln P(0) = \int$ ....

6. The integral of the relative growth rate of a population is 0.3. That means the population has increased by a factor of ....

Take note

- The absolute growth rate is constant for linear growth and the relative growth rate is constant for exponential growth.

- Remember that absolute growth rate (of a population) is measured in people per year and relative growth rate is measured in percentage per year.

Checking examples

- **Example 1:** Notice that unless the initial population size is given, we cannot find the total change in the population size. We can find the factor by which the population size has changed and so the percentage increase. Don't rush this example.

- **Example 2:** Note that because the graphs show *relative* birth and death rates, the shaded area in Fig 6.27 will give the total change in $\ln P$ and not in P.

Problems

1, 2, 3, 6, 7, 8

REVIEW OF CHAPTER 6

1. The average value of f over $0 \le t \le 10$ is 15 from which we can conclude that $\int_0^{10} f(x)\,dx =$ ......

2. The equilibrium point (q^*, p^*) of a supply and demand situation is $(500, 10)$.

   - The consumer surplus is given by the area ....
     ....

- If this expression has a value of $1000, it means that ....

....

- The producer surplus is given by the area ....

....

- If this expression has a value of $1500, it means that ....

....

3. How much should you invest now at 8% to have $1000 in 5 years?

....

4. A constant income stream of $S(t) = \$1000$ is given per year. How much money will you have in 10 years if the investment rate is 6%. ....

5.   - The integral of the (absolute) rate of change of a population over a three year period is 20. What does that tell you? ....

....

- The integral of the relative growth rate over a three year period is 0.2. By what percentage did the population increase over this period? ....

....

Review Problems

1, 4, 12, 17, 24

CHAPTER 7

ANTIDERIVATIVES

7.1   CONSTRUCTING ANTIDERIVATIVES ANALYTICALLY

CAREFULLY READ THROUGH THIS SECTION IN THE TEXTBOOK

Key points

- Explaining what an antiderivative is.

- Looking at formulas for antiderivatives.

- Looking at properties of antiderivatives.

Knowing the Book

1. .... is an antiderivative of $2x$.

   .... is the most general antiderivative of $2x$.

2. $F(x)$ is the antiderivative of $f(x)$ if ....

3. If $F(x)$ is an antiderivative of $f(x)$ then

$$\int \text{.............................} = \text{...}$$

4.   • An antiderivative of x^n is .... and therefore

$$\int x^n dx = \text{..................................}$$

   • An antiderivative of k is .... and therefore

$$\int k dx = \text{..................................}$$

   • An antiderivative of $\frac{1}{x}$ is .... and therefore

$$\int \frac{1}{x} dx = \text{..................................}$$

   • An antiderivative of e^{kx} is .... and therefore

$$\int e^{kx} dx = \text{...}$$

   • An antiderivative of $\sin x$ is .... and therefore

$$\int \sin x dx = \text{...}$$

   • An antiderivative of $\cos x$ is .... and therefore

$$\int \cos x dx = \text{...}$$

Checking the concepts

1. True/False?

   (a) An antiderivative of $10x$ is $5x^2$.   T/F

   (Is $\dfrac{d}{dx}5x^2 = 10x$?)

   (b) An antiderivative of $1/x$ is $\ln x$.   T/F

   (Is $\dfrac{d}{dx}\ldots\ldots\ldots = \ldots\ldots\ldots$?)

   (c) An antiderivative of $2x^4$ is $8x^3$.   T/F

   (Is $\dfrac{d}{dx}\ldots\ldots\ldots = \ldots\ldots\ldots$?)

   (d) An antiderivative of e^{2x} is e^{x^2}.   T/F

   (Is $\dfrac{d}{dx}\ldots\ldots\ldots = \ldots\ldots\ldots$?)

   (e) An antiderivative of e^{-2x} is $-\frac{1}{2}e^{-2x}$.   T/F

   (Is $\dfrac{d}{dx}\ldots\ldots\ldots = \ldots\ldots\ldots$?)

2. True/False?

   (a) $\int te^t dt = \dfrac{t^2}{2}e^t + C$.   T/F

   (b) $\int 3x^2 dx = x^3 + C$.   T/F

   (c) $\int \ln x\, dx = \dfrac{1}{x} + C$.   T/F

   (d) $\int \dfrac{1}{\sqrt{x}} dx = 2\sqrt{x} + C$.   T/F

Take note

Unfortunately not every function has an antiderivative that can be written using our basic functions. For example, the antiderivative of $f(x) = e^{x^2}$ cannot be written in terms of any of the functions we know. The good news is that all of the functions discussed in Chapter 1 have antiderivatives that can be easily written down.

Checking examples

These examples serve only as an appetizer and are nowhere near enough for you to become proficient in finding antiderivatives. Practice, practice and more practice is necessary.

Problems

Do all odd numbered problems, more if necessary.

7.2   INTEGRATION BY SUBSTITUTION

CAREFULLY READ THROUGH THIS SECTION IN THE TEXTBOOK

Key points

- Recognizing when to make a substitution in an integral.

- Learning how to make a substitution.

Knowing the Book

1. The chain rule is applied to $f(g(x))$ where f is the ....-side function and g is the ....-side function.
   The chain rule says: $\dfrac{d}{dx}(f(g(x))) = $ ....

2. If w is the "inside function", then $dw = $ ....dx

Checking the concepts

1. Give the substitution for each of the integrals:

   $\displaystyle \int x\sqrt{x^2+1}\,dx$ ....

   $\displaystyle \int x^2 e^{x^3}\,dx$ ....

   $\displaystyle \int x(x^2+1)^5\,dx$ ....

   $\displaystyle \int (3-t)^5\,dt$ ....

2. Which of the following can be determined by using a substitution?

   (a) $\displaystyle \int x^2\sqrt{x^2+1}\,dx$

   (b) $\displaystyle \int x^2\sqrt{x^3+1}\,dx$

   (c) $\displaystyle \int e^{\sin x}\,dx$

   (d) $\displaystyle \int \cos x e^{\sin x}\,dx$

   (e) $\displaystyle \int \frac{x}{x^2+1}\,dx$

   (f) $\displaystyle \int \frac{x^2}{x^2+1}\,dx$

3. For the substitution:

   $w = x^2+1 \quad dw = $ ....dx

   $w = 3-t \quad dw = $ ....dt

   $w = \cos 2x \quad dw = $ ....dx

Take note

Once you are comfortable with substitutions you can cut down on the writing and do it mentally.

Checking examples

- **Examples:** These examples serve to illustrate when and how to use substitution. Follow these by doing a good number of substitution exercises yourself.

Problems

Do all odd numbered problems, more if necessary.

7.3   USING THE FUNDAMENTAL THEOREM TO FIND DEFINITE INTEGRALS

CAREFULLY READ THROUGH THIS SECTION IN THE TEXTBOOK

Key points

- Using the Fundamental Theorem to get an exact answer to an integral.

- Computing a definite integral by substitution.

Knowing the Book

1. To find $\int_a^b F'(x)dx$ we first try to find .... and then calculate .... - ....

2. When computing a definite integral by substitution there are two ways of doing it. Are you familiar with both methods?

3. According to Problem 36: To calculate the improper integral $\int_0^\infty f(x)dx$ we would first calculate

   .... and then let ....

Checking the concepts

1. Given a function, find the antiderivative and determine the definite integral:

Function	Antiderivative	Definite integral
$10e^{0.01t}$	....	$\int_0^1 f(t)dt = $ ....
$3\sin 2t$	....	$\int_0^\pi f(t)dt = $ ....
x^3	....	$\int_0^4 f(x)dx = $ ....
$\dfrac{2}{x}$	....	$\int_1^2 f(x)dx = $ ....
x^{-3}	....	$\int_1^2 f(x)dx = $ ....

2. If we know that $\int_0^b e^{-0.1t}dt = -10(e^{-0.1b} - 1)$ then $\int_0^\infty e^{-0.1t}dt = $ ....

Take note

The question that you have to ask yourself here is: *If the derivative is given, what is the function?* If this question can be answered (which is not always the case) this method is quick and exact.

Checking examples

- **Example 1:** An antiderivative normally comes with an arbitrary constant C but when using the Fundamental Theorem C cancels out in the process. When using the Fundamental Theorem the arbitrary constant C can be omitted.

- **Examples 2, 3 & 4:** These are all simple but important examples to illustrate the methods.

Problems

1, 5, 9, 18, 26, 31, 36

7.4   ANALYZING ANTIDERIVATIVES GRAPHICALLY AND NUMERICALLY

CAREFULLY READ THROUGH THIS SECTION IN THE TEXTBOOK

Key points

- Calculating $F(b)$ when $F(a)$ and F' is given.

- Graphing a function given the graph of the derivative (with certain areas indicated).

Knowing the Book

The Fundamental Theorem is used here in the form

$$F(b) = F(a) + \text{..........................}$$

Checking the concepts

1. The graphical relationship between the derivative and the function is emphasized again in this section. Match the following which all refer to features of the graph of F':

1.	F' lies above the t-axis	A.	F decreases
2.	F' cuts the t-axis from above	B.	F has an inflection point
3.	F' has a turning point	C.	F increases
4.	F' changes sign from negative to positive	D.	F has a local maximum
5.	F' lies below the t-axis	E.	F has a local minimum

2. If the graph of F' is positive and increasing then the graph of F ....-creases at a(n) .....-creasing rate.

   If the graph of F' is positive but decreasing then the graph of F ....-creases at a(n) .....-creasing rate.

3. If $\int_0^2 f'(t)dt = 4$ and $F(0) = 3$ then $F(2) =$ ....

   If furthermore $\int_2^4 f'(t)dt = -3$ then $F(4) =$ ....

Checking examples

- **Example 2:** The change in F over an interval is the area below the curve F' on that interval, so $F(b) = F(a) +$ area below F' between a and b.

- **Example 3:** This is an important example. Where the rate of change graph turns below the x-axis, where $t = 2$, is where the concentration graph decreases most rapidly. Where the rate of change graph turns above the x-axis, at $x \simeq 6.7$, is where the concentration graph increases most rapidly.

- **Example 4:** Note again that there are many functions with this particular derivative but as soon as $f(0) = 10$ is specified then f is unique.

Problems

1, 3, 4, 8-9, 12,

REVIEW OF CHAPTER 7

1. The antiderivatives of the following functions are:

   $3x + 1$ ....

   $4 \sin 3t$ ....

   e^{t+4} ....

2. Give the substitution required to determine the following:

   $$\int \frac{e^{\sqrt{x}}}{\sqrt{x}} dx = \qquad \text{.................}$$

   $$\int (4 + \cos\theta)^4 \sin\theta \, d\theta = \qquad \text{.................}$$

   $$\int \frac{x}{x^2 + 4} dx = \qquad \text{.................}$$

3. $\displaystyle\int_0^4 e^{0.1t} = f(\ldots) - f(\ldots)$ where $f = $ ....

4. Describe the behavior of the antiderivative if:

   - The function is positive and decreasing.

     ....

   - The function is constant.

     ....

   - The function cuts the x-axis from below.

     ....

Review Problems

1, 5, 10, 15, 21, 26, 32, 36, 40, 47, 52

CHAPTER 8

PROBABILITY

8.1 DENSITY FUNCTIONS

CAREFULLY READ THROUGH THIS SECTION IN THE TEXTBOOK

Key points

- Using age distribution in the US to illustrate what a density function is.

- Looking at properties of a density function.

Knowing the Book

1. For the age density function the units on the y-axis are

   ....

   and the units on the x-axis are ....

2. The area of each bar in the histogram in Fig 8.1 represents the ....
   in that age group. The total area of all the rectangles is has a value of ......

3. By "smoothing out" such a histogram we get a ....

4. For the density function $p(x)$:
   $$\int_a^b p(x)dx \text{ represents}$$
   ....

5. List two characteristics of a density function $p(x)$:

   - ....

   - ....

6. We do not interpret $p(10)$ (for example), but rather $p(10)\Delta x$. How is this
   interpreted? ....
   ....

Checking the concepts

1. For the age density function discussed in this section, what does $\int_{20}^{30} p(x)dx$
   represent? ....
   ....

2. If the distribution of the height (in inches) of adult females in a population is studied and the density function $p(x)$ is constructed:

The x-axis will represent ....

The y-axis will represent ....

The area under the curve between two x-values represents ....

....

3. For the height example above, what does $\int_{60}^{62} p(x)\,dx$ represent? ....

....

Take note

A density function should be seen as a derivative.

Checking examples

- **Example 1:** This example illustrates two points:

  - the fact that we always work with areas and never with function values when dealing with a density function.

  - the fact that ages in any age group are not uniformly distributed (look especially at the 80 - 100 group).

- **Example 2:** What fraction of patients will wait for less than half an hour? What fraction of patients will wait for more than 2 hours?

Problems

1, 2, 5, 6, 13, 16

8.2   CUMULATIVE DISTRIBUTION FUNCTIONS AND PROBABILITY

CAREFULLY READ THROUGH THIS SECTION IN THE TEXTBOOK

Key points

- Defining the cumulative distribution function.

- Using a density function to define probability.

- Introducing the normal distribution.

Knowing the Book

1. The cumulative distribution function $P(t)$ for a density function $p(t)$ is defined by

$$P(t) = \int_{....}^{....} \text{...}$$

2. List three features of a cumulative distribution function:

   - ....

   - ....

   - ....

3. When dealing with a *density function* then the fraction of a population between a and b is given by:

   ....

   but when working with a cumulative distribution function the fraction of a population between a and b is given by:

   ....

Checking the concepts

1. What does $P(6)$ represent if $p(t)$ is a age density function? ....
   ....

2. If $p(t)$ is the density function and $P(t)$ is the cumulative distribution function, then $\int_{3}^{5} p(t)dt = P(...) - P(...)$

3. If $p(t) = 0.1$ for $0 \le t \le 10$:

   - Sketch $p(t)$ and $P(t)$

   - As a formula: $P(t) = $ ....

   - $P(0) = $ ....

- $P(10) = $ .... and so $\int_0^{10} p(t)dt = $ ....

- $\int_2^4 p(t)dt = $ ....

4. If $h(t)$ is a height distribution then the probability that a person is taller than 60 inches is given by: ....

Checking examples

- **Example 1:** This example emphasizes the difference and similarities between a density function and a cumulative distribution function. It is a lengthy example but illustrates the concepts well. Spend enough time studying it.

- **Example 2:** Notice that the density function should be seen as a derivative. Because the density function is positive (above the x-axis) the cumulative distribution is increasing. Don't be misled by the fact that the density function increases at first and then decreases. It simply means that the cumulative function increases at an increasing rate at first and then it increases at a decreasing rate.

Problems

2, 3, 9, 118

8.3 THE MEDIAN AND THE MEAN

CAREFULLY READ THROUGH THIS SECTION IN THE TEXTBOOK

Key point

- Calculating the median and the mean of a feature (such as age) of a population when given a density function.

Knowing the Book

1. The **median** of a quantity x distributed through a population is a value T such that .... of the population has values of x less than .... and half the population has values of x .... than T.

2. When given the density function $p(t)$, the median can be calculated by solving for T from the equation

....

3. When given the cumulative distribution function $P(t)$ we look for a number T for which:

....

4. The mean value can be calculated from the density function $p(t)$ by calculating the definite integral

....

5. What is the meaning of the standard deviation of the normal distribution? ..

....

Checking the concepts

1. If five students scored as follows: 1, 4, 5, 6, 6 then the median is: .... and the mean is ....

2. Is the median T for a density function a value on the horizontal axis or on the vertical axis? ....

3. When finding the *median* we look for a point on the horizontal axis which divides the area below the density function in ....

4. The *mean* is the point on the x-axis where the graph of the density function (made from cardboard) would ....

Take note

1. The *mean* value formula might seem strange at first. It seems different from our usual concept of average. The following simple discussion might provide the missing link: The ten numbers 4, 4, 4, 8, 8, 8, 8, 8, 8, 8 have an average of 6.8 (add up and divide by 10) The average could also have been calculated as follows:

$\frac{1}{10}(3 \times 4 + 7 \times 8)$

which is the same as

$\frac{3}{10} \times 4 + \frac{7}{10} \times 8$

We now have

fraction of 4's in group $\times$ 4 + fraction of 8's in group $\times$ 8

This expression is very similar to the integral definition of the mean. In the continuous case the sum becomes an integral and the fractions become a density function.

2. The formula for the normal distribution is rarely used. Tables with values of the standard normal distribution are freely available.

Checking examples

- **Example 1:** Note that the area below the graph is 1, so it is indeed a density function. For the (c) part, the Fundamental Theorem could also have been used. In this case:

$$\int_0^T p(t)dt = (0.04t - 0.0004t^2)|_0^T = 0.5$$

  Solving for T from $0.0004T^2 - 0.04T + 0.5 = 0$ renders $T = 14.6$ or $T = 85.35$. The value $T = 14.6$ makes sense but where does $T = 85.35$ come from?

- **Example 2:** This integral could also easily have been done using the Fundamental Theorem.

Problems

2, 3, 4, 7, 11, 14, 18

REVIEW OF CHAPTER 8

1. $f(t)$ is a density function of waiting time (in minutes) in a queue at a shop. The horizontal axis shows .... and the area under the graph tells you ....

   - $\int_3^5 f(t)dt$ represents ....

   - If $F(t)$ is the cumulative density function, then in terms of probability $F(5)$ represents ....
   ....

   - If $\int_0^3 f(t)dt = 0.5$ what can you say about the median of this distribution? ....

2. Match the following for an age density function $p(x)$:
   1. percentage between ages 40 and 50   (a) $P(50)$
   2. percentage younger than 50   (b) $\int_{-\infty}^{\infty} xp(x)dx$
   3. the median   (c) value of T such that $P(T) = 0.5$
   4. the mean age   (d) $\int_{40}^{50} p(x)dx$

Review Problems

3, 6, 8, 9

CHAPTER 9

FUNCTIONS OF SEVERAL

VARIABLES

9.1   UNDERSTANDING FUNCTIONS OF TWO VARIABLES

CAREFULLY READ THROUGH THIS SECTION IN THE TEXTBOOK

Key points

- Investigating functions of two variables numerically or algebraically.

- Investigating functions of two variables by varying one variable at a time.

Knowing the Book

1. If R is a function of x and y we write ....

2. The dependent variable(s) of $P = f(V, T)$ are:.............. and the independent variable(s) are: ....

3. The domain of a two-variable function is ....

4. If one variable is kept fixed while the other one varies, the remaining function of one-variable is called ....

Checking the concepts

1. Use appropriate variables to write the following in symbols:

   - The height of an oak tree depends on the age and the fertility of the soil.
     ....

   - Your ability to concentrate depends on the time of day and how interesting the topic is.
     ....

2. Give **any** *algebraic* example of each of the following two-variable functions:

   - $M = f(C, t) =$ ....
   - $S = f(r, g) =$ ....

3. If P is a function of the three variables L, k, and t then $P =$ ....

4. If $P = f(I, t)$ is the size of a population in millions where I represents the initial number and t time in years:

   - The number $f(2.5, 10)$ represents the ....
     ....

- The expression $f(I, 15)$ represents the ....

....

- The expression $f(4, t)$ represents the ....

....

- A possible formula for P (assuming a growth rate of 2% per year) is $f(I, t) =$ ....

Checking examples

- **Example 1:** The more general form of these two formulas would be a three-variable function where the interest rate also varies. For example, for continuous compounding the formula then becomes $M = f(B, t, r) = Be^{rt}$

- **Example 2:** This two-variable function is the equivalent of the linear functions from Chapter 1.

- **Examples 3 & 4:** It is not easy to visualize two-variable functions but it becomes easier with time and practice.
  Note that if the x-variable is fixed, we are left with the well-known surge function. If the t-variable is fixed, on the other hand, we are left with an exponential function. By using software that can sketch two-variable functions we can get a better over-all idea of what the function looks like.

Problems

1 - 5, 7, 13

9.2   CONTOUR DIAGRAMS

CAREFULLY READ THROUGH THIS SECTION IN THE TEXTBOOK

Key points

- Illustrating what a contour diagram is.

- Using a contour diagram.

- Making a contour diagram from a table.

- Finding contours algebraically.

Knowing the Book

1. Curves for which $f(x, y)$ is .... are called contours.

2. Every point along the same contour has the same ....

3. For the topographical map: What are contour curves called? ....
   The more closely spaced the contour lines, the .... the terrain.

4. If we have a formula for $f(x, y)$, the equation for the contour of value c is given by

   ....

5. If $f(x, y) = z$ then the contour lines are usually drawn for equally spaced values of ....

Checking the concepts

1. Two contours can never cross. Can they ever touch? ....

2. What does it mean if the temperature contours are widely spaced on a weather map?
   ....

3. What shape do the contours of the function $f(V, T) = \dfrac{2V}{T}$ have?
   ....

Checking examples

- **Example 1:** This example serves to familiarize yourself with a contour diagram. It also shows how to estimate the temperature for towns between isotherms.

- **Example 2:** This example shows the contour map and a 3-D representation of the function.

- **Example 3 & 4:** An important example. Two skills are illustrated here:
  - Reading from a contour diagram.
  - Moving along a horizontal line or a vertical line and interpreting the behavior of the contours.

- **Example 5:** It is important to understand what the contour diagram tells us. Remember that a high GFR indicates healthy kidneys and a low GFR indicates ill kidneys.
  Do not rush these examples. The skills of the previous example are illustrated once again.

- **Example 6:** Drawing contours from a table of values is confusing because you have to think "upside-down". The humidity values in the table increase downwards but the same values on the graph increase upwards. Check your understanding by drawing a similar contour for a heat index of 100 on *Figure 9.17*. It will be wise to plot a few points and fit a hand-drawn curve through the points. (Take values of the heat index between 99 and 101 as 100.)

- **Example 7:** This example illustrates the method for constructing contours algebraically. When a two-variable function is given algebraically, set the function equal to some constant for sketching contours. Find a relationship between the two independent variables and you'll probably find a well-known curve such as a line or a circle.

Problems

2, 3, 7, 12, 18, 25, 28

9.3   PARTIAL DERIVATIVES

CAREFULLY READ THROUGH THIS SECTION IN THE TEXTBOOK

Key points

- Defining partial derivatives.

- Estimating partial derivatives from:
  - a table.
  - a contour diagram.

- Using a partial derivative to estimate function values.

Knowing the Book

1. The partial derivatives $f_x(x, y)$ and $f_y(x, y)$ are defined as follows:

   $f_x(a, b) =$ ....

   $f_y(a, b) =$ ....

2. Alternative notation if $z = f(x, y)$:

   $\dfrac{\partial z}{\partial x} =$ ....

   $\dfrac{\partial z}{\partial y} =$ ....

3. For a contour diagram: If the values on the contours are increasing as we move parallel to one of the axes, then the partial derivative in that direction is ....-tive.

4. The relationship of local linearity: $\Delta f =$ .... + ...

Checking the concepts

1. The partial derivatives of the function $f(s, g)$ are .... and ....

2. If $f(B, t)$ is the amount of money in a bank account, t years after an initial investment of B dollars, then:

   If $f_B(1000, 2) = 2.50$, it tells you that if you invest \$1 more than \$1000, then after 2 years you will have approximately \$..............
   If $f_t(1000, 2) = 30.50$, it tells you that ....
   ....

3. If $f(10, 5) = 30$, $f_x(10, 5) = -1.5$ and $f_y(10, 5) = 3.5$ then

   - $f(11, 5) \simeq$ ....
   - $f(10, 7) \simeq$ ....
   - $f(12, 6) \simeq$ ....

Checking examples

- **Example 1:** Note that the calculated partial derivative is constant for all values because of the nature of the function.

- **Example 2:** As in the one dimensional case a partial derivative can be estimated from a table in one of three ways: moving forward, moving backward or taking the average of the two (normally more accurate). For two variables "moving forward" means moving to the right in the table (in the horizontal direction) or moving down in the table (in the vertical direction). "Moving backwards" means moving left or up in the table.

- **Example 3:** This example is followed by the local linearity formula which really is only an extension of the one dimensional case. A very useful formula.

- **Example 4:** Check whether you know how partial derivatives are calculated from a contour diagram:

  - Estimate $H_x(20, 10)$ and interpret your answer.

  - Estimate $H_t(20, 10)$ and interpret your answer.

Problems

1, 2, 3, 6, 7, 10, 24

9.4   COMPUTING PARTIAL DERIVATIVES ALGEBRAICALLY

CAREFULLY READ THROUGH THIS SECTION IN THE TEXTBOOK

Key points

- Computing partial derivatives algebraically.

- Introducing Cobb-Douglas Production Functions.

- Defining Second-Order Partial Derivatives.

Knowing the Book

1. The partial derivative $f_x(x,y)$ is the ordinary derivative of the function $f(x,y)$ with respect to ...., with y fixed

   and:

   The partial derivative $f_y(x,y)$ is the ordinary derivative of the function $f(x,y)$ with respect to ...., with x fixed.

2. For fixed N (number of workers) and increasing value of equipment V, one would want production P to increase at a ....-creasing rate.

3. What does the Cobb-Douglas function model? ....
   General formula:

   ....

   where ....

4. List the four types of second-order partial derivatives for the function $f(x,y)$:

   ....

   ....

   Which two are the same?

Checking the concepts

True/False?

1. (a) If $f_x(x,y) = 0$ then $f(x,y) = C$.   T/F

   (b) If $f(x,y) = e^{xy}$ then $f_y(x,y) = xe^{xy}$.   T/F

   (c) If $f(x,y) = \frac{x}{y}$ then $f_y(x,y) = \frac{-x}{y^2}$.   T/F

   (d) $f(x,y) = 10L^3K^4$ is a Cobb-Douglas function.   T/F

2. For a Cobb-Douglas function $P = f(N,V)$ (P the number of cars produced per month, N the number of workers and V the value of the equipment in millions of dollars), what does the following tell you?

   - $f_N(50,100) = 2$ ....
     ....
   - $f_V(50,100) = 20$ ....
     ....

Checking examples

- **Example 1:** The technique for calculating partial derivatives algebraically is set out in this example. Note that the derivatives are calculated first and then values are substituted.

- **Example 2 & 3:** It takes a bit of practice to think of one variable as a constant when calculating partial derivatives, but working through the functions in this example will secure the technique.

- **Example 4:** We see that in this example, $f_x(1,2) = -0.5413$. What does it mean?

  ....
  ....

  We also see that $f_y(1,1) = 0$. What does that mean? ....
  ....
  ....

  Show these two slope values on the graphs.

- **Example 5:** Show the following statement graphically on the contour diagram in *Figure 9.47*:

  Fixing V at a particular value and letting N increase, you eventually find yourself moving nearly parallel to the contours, crossing them less and less frequently.

- **Example 6:** Interpretation of derivatives is what this example is all about.

- **Example 7:** When using a table of values it is necessary to compute the first derivative f_x at two different (but neighbouring) points in order to calculate the second order derivative.

- **Example 8:** Calculating a second order derivative is simply repeating the calculation of a first order derivative.

Problems

1, 6, 7, 13, 18, 21, 25, 34

9.5 CRITICAL POINTS AND OPTIMIZATION

CAREFULLY READ THROUGH THIS SECTION IN THE TEXTBOOK

Key points

- Defining local and global maxima and minima for functions of two variables.

- Finding a local maximum or minimum analytically.

- Deciding whether a function has a local maximum or a local minimum at a critical point.

Knowing the Book

1. f has a local minimum at P_0 if .... for all points P near ....
   For f to have a global minimum at P_0, $f(P_0) \geq f(P)$ for all points P ....

2. f has a local maximum at P_0 if .... for all points P near ....
   For f to have a global maximum at P_0, $f(P_0) \geq f(P)$ for all points P ....
   Note that a *point* P_0 has two coordinates and is of the form (x_0, y_0).

3. For both one-variable and two-variable functions:
   When looking for extrema the first thing to do is to find the ....
   points.

4. *Critical points* (not on the boundary of the domain) of a two-variable function
   are where .... **and** ....
   or ....

5. In order to use the Second Derivative Test at a critical point (x_0, y_0) where
   both f_x and f_y are zero, we first have to work out the number

   $$D = \text{...}$$

   The sign of D and the sign of $f_{xx}(x_0, y_0)$ will determine whether we have a
   local maximum, local minimum or neither.

6. If the point (x_0, y_0) is a critical point, does it mean that f has a local maximum
   or a local minimum in the point? ....

Checking the concepts

1. For $f(x,y) = x^2 - y^2$ we see that $f_x(0,1) = 0$. Does it follow that (0,1) is a critical point? ....
....

2. Is (1,2) a critical point of $f(s,t) = s^2 - st + t$? ....
....

3. The function f has no critical points. Is it possible that it can have a global minimum nevertheless? ....

4. True/False?

    (a) A local minimum can only occur at a critical point.   T/F

    (b) A global maximum can only occur at a critical point.   T/F

    (c) Every function has a global maximum.   T/F

5. Why would you say the production function $f(N,V) = 2N^{0.6}V^{0.4}$ has a global minimum at (0,0)? ....
....

Take note

It is necessary to draw up a table in order to establish whether a function has a local extreme at a critical point **unless** the second derivative test is used. Remember that f_x as well as f_y can be zero at a point without it being a local minimum or local maximum. An example of this is a pass through the mountains - see Figure 9.10 on p 328.

Checking examples

- **Example 1:** In this case the global maximum occurs at a local maximum but it could also have occurred at a boundary.

- **Example 2:** The local maximum appears to be a global maximum as well. Should we not have looked on the boundaries as well for the global maximum?

- **Example 3:** See hint.

- **Example 4:** This example covers all the principles in this section and revisits cost, revenue and profit. Study it well and don't forget to brush up on your simultaneous equations!

- **Example 5:** The second derivative test is often useful and it saves drawing up a table of values around the critical point.

Problems

1, 2, 5, 8, 12, 18

9.6   CONSTRAINED OPTIMIZATION

CAREFULLY READ THROUGH THIS SECTION IN THE TEXTBOOK

Key points

- Introducing constrained optimization.

- Using the Method of Lagrange Multipliers.

- Looking at the meaning of λ.

- Using a Lagrangian function.

Knowing the Book

1. For the budget constraint line in *Figure 9.53*:
   Any point below the line represents values of x and y that are .... (within/outside) the budget.
   Any point above the line represents values of x and y that are .... (within/outside) the budget.
   Any point on the line represents values of x and y that ....the budget.

2. The optimum point in this example occurs where the budget constraint line is .... to a contour.

3. In general: If $f(x, y)$ has a global maximum or minimum on the constraint $g(x, y) = c$, it occurs at a point where
   ....
   or at ....

4. Suppose P_0 is a point on the constraint $g(x, y) = c$.

   - When does f have a local minimum (not maximum) at P_0 subject to the constraint?
   ....

- When does f have a global minimum at P_0 subject to the constraint?

  ....

5. Make sure that you know how to apply the Method of Lagrange Multipliers.

6. For the production problem the value of λ represents the extra ....

   .... achieved by increasing the .... by 1.

   (or the extra .... for the extra ....)

7. Interpreting the Lagrange multiplier in general:

   The value of λ is approximately the increase in the ....

   .... when the value of the ....

   is increased by 1.

   The value of λ represents the ....

   of the optimum values of f as the constraint increases by 1.

8. How is the Lagrangian function defined? ....

Checking the concepts

1. Draw up a budget constraint function if one item costs \$20 and another costs \$15 and you're not allowed to spend more than \$500.

   ....

2. For a production function f the budget constraint is $10x + 10y \leq 2500$.

   - Sketch the budget constraint.

   - Purchasing the two quantities $x = 100$ and $y = 100$ costs ....
     This .... (is/isn't) within the budget. Plot this point. It lies
     .... the budget constraint.

- Purchasing the two quantities $x = 200$ and $y = 200$ costs ....
This .... (is/isn't) within the budget. Plot this point. It lies
.... the budget constraint.

- A point that satisfies the budget exactly is the point ...., for
example. The point lies .... the budget constraint.

3. If the budget constraint function is tangent to a production contour $P = 1300$
at $(12, 13)$ it means that ....
....
....

4. If $\lambda = 0.034$ it means that if the budget is increased by ....
then production will increase by ....

Checking examples

- **Example 1, 2, & 3:** The examples are all related and form the backbone of
this section. The concepts discussed above are all developed in these examples.

- **Example 4:** This example does not introduce new concepts, it only illustrates
the same concepts with a different production function and different constraint.

Problems

3, 9, 11, 15, 18

REVIEW OF CHAPTER 9

1. Use function notation (and your own variables) to write the following in symbols:

- The amount of food grown depends on the amount of rain and the amount
of fertilizer used.
....

- The rate of a chemical reaction depends on the temperature and the
pressure of the environment in which it proceeds.
....

- The monthly mortgage payment in dollars depends on the amount bor-
rowed, the interest rate and the number of years before the mortgage is
paid off.
....

2. The number Q of elephants, in thousands, that a game park can carry depends on the size s of the park in thousands of square miles, and rainfall r in the park in inches per year, so $Q = f(s, r)$. (The figures below relate to the Kruger National Park in South Africa.)

  (a) $f(8, 19) = 7$ means that ....
  ....

  (b) $f_x(8, 19) = 1$ means that ....
  ....

  (c) $f_y(8, 19) = 0.1$ means that ....
  ....

3. The figure below shows a contour diagram of a function $P = f(x, y)$, with steadily increasing contour lines ($P = 100$, $P = 200$ and $P = 300$ for example.)

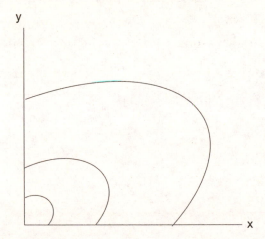

  (a) Is f_x positive or negative? ....
  (b) Is P increasing at an increasing or decreasing rate in the x-direction?
  ....

4. The function $f(x, t) = \dfrac{e^{-t}}{x}$ ....-creases in the t-direction and ....-creases in the x direction.

5. The general form of the Cobb-Douglas function is ....

6. The point $(3,4)$ is a critical point of $f(x, y)$ if
  ....

  or ....

7. One identifies a local maximum for a two-variable function by finding a critical point and then ....
....

8. For a production function $Q(x, y)$ and budget constraint $g(x, y) = c$:

    (a) x and y represent ....

    (b) Q represents ....

    (c) The optimum value of Q (if it exists) can be found by solving:

        ....

        ....

        ....

    (d) If the Lagrange multiplier λ is 0.004, it means that ....
    ....
    ....

Review Problems

1, 2, 3, 4, 11, 21, 30

FOCUS ON THEORY

DERIVING THE FORMULA FOR A REGRESSION LINE

CAREFULLY READ THROUGH THIS SECTION IN THE TEXTBOOK

Key points

This section is devoted to an explanation of how the formula for finding a regression line to a set of data is obtained. Knowledge of two-variable functions is necessary for the derivation. That is the reason why it is done only at this advanced stage in the book although regression lines were already introduced in Chapter 1. Example 1 illustrates how a particular regression line is obtained through applying the concepts introduced in Chapter 9. It is important to follow this example. What was done in numbers in Example 1 is done in symbols in the derivation on p 371. The explanation is not difficult to follow, only keep the following in mind: x_i and y_i are constants (numbers) whereas m and b are the variables with respect to which we differentiate.

Problems

1, 3, 7

CHAPTER 10

MATHEMATICAL MODELING

USING DIFFERENTIAL

EQUATIONS

10.1 MATH MODELING: SETTING UP A DIFFERENTIAL EQUATION

CAREFULLY READ THROUGH THIS SECTION IN THE TEXTBOOK

Key point

- Setting up a differential equation from a verbal description.

Knowing the Book

In this section a number of real-life situations are discussed where the end result in each case is a differential equation. Note that the left-hand side of a differential equation is always a derivative of an unknown function and the right-hand side is an expression involving the unknown function. The aim is to determine the unknown function. How to do this becomes clear in the subsequent sections. Setting up a differential equation is a skill that you need to acquire and it comes through practice. By working through these examples you become familiar with translating words into differential equations. No new mathematics here, just a new way of thinking.

Checking the concepts

1. For the differential equation $\dfrac{dP}{dt} = 0.20(P - 10)$, P in millions:

   - At what rate does the population grow/decay when there are 2 million fish (t in years)?
     ....

   - Why is it not possible to say off-hand at what rate the population will change when $t = 3$?
     ....

   - What is the population size when it grows at a rate of 0.5 million fish/year?
     ....

2. Set up a differential equation for the following simple cases:

   (a) The rate at which a population grows is directly proportional to the size of the population. ....

   (b) A population grows at a continuous rate of 5% of its size........................

   (c) A town has 10 000 people. The rate at which they become ill (due to flu) is directly proportional to the number of people not yet ill. Let P be the number of ill people. ....

   (d) Your money decreases continuously at 2% per year continuously.............

Checking examples

- **Marine Harvesting:** The fish are caught at a constant rate of 10 million fish per year. This is a bit of a difficult concept. Every year 10 million fish are caught but this happens at a constant rate throughout the year, day in, day out, every moment.

- **Net worth of a Company:** Again, money is paid out at a constant rate. So the money is paid out at the same rate continually throughout the year. This seems unlikely but is mathematically convenient.

- **Example 1:** This is one of the simplest differential equations one can get. Yet this type of differential equation has wide application and is important.

- **Example 2:** The term $-0.1Q$ means that the drug decreases at a rate of 10% of what is present.

- **The Logistic Model:** This example looks at logistic growth from another angle. This differential equation is the most involved of the section and involves the number of people (P) and the space left $(L - P)$.

Problems

1, 2, 7, 9, 10, 12, 16

10.2 SOLUTIONS OF DIFFERENTIAL EQUATIONS

CAREFULLY READ THROUGH THIS SECTION IN THE TEXTBOOK

Key points

- Looking at examples of solutions of differential equations - numerically and as a formula

- Dealing with an initial condition.

- Discussing the difference between a general and the particular solution of a differential equation.

Knowing the Book

1. The initial condition is used to select a .... solution from the family of solutions.

2. The differential equation and the initial condition together are called an ....

....

Checking the concepts

1. For the differential equation $\dfrac{dP}{dt} = 5 - 0.1P$, we know that $P = 10$ at $t = 0$. Estimate $P(1)$: ....

2. The general solution of $\dfrac{dP}{dt} = 0.1P$ is $P(t) = Ce^{0.1t}$. What is the particular solution if $P(0) = 100$?

3. For the differential equation $\dfrac{dy}{dx} = y$, which of the following is a solution:

   (a) $y = \frac{1}{2}x^2$ or

   (b) $y(x) = e^x$

Checking examples

- **Example 1:** This example shows the difference between the general and particular solutions of a differential equation. There is only *one* particular solution but the general solution is a family of solutions.

- **Example 2:** Is $y = e^{2x}$ a solution of the differential equation?

- **Example 3:** Note what the differential equation says, translated into words: y decreases continuously by 50% of the value of y. This is the reason why the solution contains the term $e^{-0.50t}$ which indicates a continuous decrease of 50% per time unit.

2, 7, 9, 10, 16

10.3   SLOPE FIELDS

CAREFULLY READ THROUGH THIS SECTION IN THE TEXTBOOK

Key points

- Visualizing the solution of a differential equation by using of a slope field.

- Touching existence and uniqueness of solutions.

Knowing the Book

1. At any one point the slope field shows the .... in which the curve goes from there.

2. If you start at a particular point on a slope field and trace the solution, then there will only be .... such a solution, so the solution is ....

Checking the concepts

1. If $\dfrac{dy}{dx} = x - y$, what is the slope of the solution at $(2, 3)$?

2. For the differential equation $\dfrac{dy}{dx} = x - y$, what behavior does the slope field show on the line $y = x - 1$? ....

3. The right-hand side of $\dfrac{dy}{dx} = y^2 + 1$ is a function of y (no $x's$). All the slope lines will be the same on lines parallel to the ....-axis.

   The right-hand side of $\dfrac{dy}{dx} = x^3$ is a function of x (no $y's$). All the slope lines will be the same on lines parallel to the ....-axis.

4. Match the differential equation and the property of the slope field:

   1. $\dfrac{dy}{dx} = y^2 + 1$ A.   all slopes are negative for $x < 0$ and $y < 0$.

   2. $\dfrac{dy}{dx} = 2 - y$ B.   slopes are positive for $y < 2$ and negative for $y > 2$

   3. $\dfrac{dy}{dx} = x + y$ C.   all slopes are positive

Take note

Move on the slope field so that the slope lines are tangent to your path.

Checking examples

In all of these examples look for features such as positive slopes or special behavior patterns. Study these well in order to get a feel for differential equations and their solutions.

Problems

1, 3, 4, 6, 14

10.4 EXPONENTIAL GROWTH AND DECAY

CAREFULLY READ THROUGH THIS SECTION IN THE TEXTBOOK

Key points

- Looking at the general solution to the differential equation $\frac{dy}{dx} = y$.

  - Looking at four applications of this differential equation:

    - Population growth.

    - Continuously compounded interest.

    - Pollution in great lakes.

    - Quantity of drug in a body.

Knowing the Book

1. The general solution to the differential equation $\frac{dy}{dt} = y$ is given by:

   ....

2. The general solution to the differential equation $\frac{dy}{dt} = ky$ is given by:

   ....

3. For $k < 0$ the solution of $\frac{dy}{dt} = ky$ is an exponentially ....-creasing function.

4. The constant C in the general solution of $\frac{dy}{dt} = ky$ is the value of y when $t = $ ....

Checking the concepts

1. The differential equation $\frac{dy}{dt} = 0.1y$ refers to a population that continually .....-creases at a rate of ....% per year.

2. The differential equation $\frac{dy}{dt} = 2y$ refers to a population that continually .....-creases at a rate of ....% per year.

3. The general solution of $\frac{dQ}{dt} = -0.05Q$ is ....
   If $Q(0) = 250$ the particular solution is ....

4. Give the differential equation for:

- A population that decreases continuously at 10% per year ....

- Money that grows at 2.5% per year, continuously compounded. ....

5. Pollution in the water:

   If Q is the quantity of pollution (in kg) and V the volume of water then $\dfrac{Q}{V}$ is the .... of pollution.

   r is the outflow rate.

   $\dfrac{rQ}{V}$ is the .... of pollution per unit .... that leaves the water.

6. Quantity of a drug in the body:

   Write in symbols: The rate at which the drug leaves the body is proportional to the quantity of the drug left in the body.

   ....

Checking examples

- **Example 3:** Note the units of $\dfrac{r}{V}$: The rate r is given in km^3/year and the volume V is given in thousands of km^3. The number value of V in the table, 0.46, is multiplied by 1000 before it can be used.

- **Example 4:** Note once again that it is not necessary to have the initial value Q_0 if we're dealing with percentage decay. Q_0 will cancel out.

Problems

1, 4, 7, 8, 10, 13, 17

10.5   APPLICATIONS AND MODELING

CAREFULLY READ THROUGH THIS SECTION IN THE TEXTBOOK

Key points

- Introducing the differential equation $\dfrac{dy}{dt} = k(y - A)$ and its applications.

- Looking at stable and unstable equilibrium solutions.

Knowing the Book

1. The general solution of $\frac{dy}{dt} = k(y - A)$ is

   ....

2. One finds equilibrium solutions by setting the derivative function equal to .....

3. What is the equilibrium solution of $\frac{dy}{dt} = k(y - A)$? Is it stable?

   ....

4. Newton's Law of Heating and Cooling:

   The temperature of a hot object decreases at a rate proportional to the difference between its temperature and the temperature of ....

Checking the concepts

1. The general solution to $\dfrac{dy}{dt} = 2(y - 10)$ is ....
   If $y(0) = 6$ the particular solution is ....

2. The equilibrium solution for $\dfrac{dy}{dt} = 2(y - 10)$ is ....

3. Which will cool down at the highest rate initially: A cup of coffee of 180° F or a cup of coffee of 100° F, if both were left to cool down? ....

4. The differential equation $\dfrac{dy}{dt} = 0.1(y - 70)$ describes the cooling rate of a cup of coffee with temperature in °F and time in minutes.

   - The surrounding temperature is ....

   - When the coffee has a temperature of 200°F then it cools down at a rate of ....

   - When the coffee has a temperature of 150°F then it cools down at a rate of ....

   - The general solution of the equation is ....

   - The initial temperature of the coffee was 200°F which means that the particular solution is ....

   - After 10 minutes the temperature of the coffee was ....

   - The equilibrium solution of the differential equation is ....
     Is the equilibrium solution stable? ....

5. What happens to the term $e^{-0.001t}$ if t becomes very large? ....

Checking examples

   - **Example 1:** Practice exercises.

- **Examples 2 & 3:** These examples refer back to Section 10.1 where the differential equations were constructed. A hint is that the value of C is always given by *Initial value - A*.

- **Example 4:** Graphically it is easy to see whether an equilibrium is stable. Alternately, one can look at the constant in the power of e in the solution. If it is positive (See Figure 10.35) then the solution is unstable, if it is positive (See Figure 10.34) then the solution is stable.

- **Example 5:** Take note of the short way of writing Newton's Law of Cooling. Also notice that the constant k is always positive and that the general solution contains two unknown constants, C and k. We need the initial temperature and the temperature at one other time for obtaining values for these contants.

Take note

When solving differential equations based on Newton's Cooling Law: Remember to factor the right-hand side (if it hasn't been done already) in order to have the function (W or y or P) "clean". For example $500 - 0.1W = -0.1(W - 5000)$

Problems

1, 6, 8, 11, 12, 14, 19, 25

10.6   MODELING THE INTERACTION OF TWO POPULATIONS

CAREFULLY READ THROUGH THIS SECTION IN THE TEXTBOOK

Key point

Constructing the Predator-Prey Model.

Knowing the Book

According to the description of the worm-robin interaction:

1. If there were no worms then the robin population will ....-crease according to the differential equation ....

2. If there were no robins the worm population will ....-crease according to the differential equation ....

3. The term describing the number of encounters is ....

4. On the phase plane .... is plotted against ....

5. What are the equilibrium points for w and r?

....

Checking the concepts

1. If there were no encounters, a robin population will decrease at 2% per year and the worm population will grow at 50% per year. The two differential equations that describe this situation are:

....

....

How do these differential equations change if there are encounters between the two populations?

2. Say there were very few robins and 1 million worms initially, show on the phase plane in Figure 10.40 what the outcome will be.

3. How many are there of the following?

   • Trajectories for a predator prey model.
   • Solutions for a predator prey model.

4. What is the difference between a trajectory and the solutions to the two differential equations?

....

Checking examples

   • **Example 1:** Follow this example up with problems 8 - 10 to enhance your understanding.

Problems

1 - 4, 8 - 14

10.7   MODELING THE SPREAD OF A DISEASE

CAREFULLY READ THROUGH THIS SECTION IN THE TEXTBOOK

Key point

Modeling the spread of flu in a boarding school.

Knowing the Book

1. We assume that the rate at which susceptible people become infected is proportional to ....

2. The number of contacts between the susceptibles S and the infected I is proportional to ....

3. We assume that the recovery rate of the infected is proportional to ....
....

4. Will an epidemic occur if the initial number of susceptibles is below the threshold value? ....

Checking the concepts

1. The number of *susceptibles* ....-creases with time. Similarly, the number of *recovered* ....-creases with time. The number of *infected*, on the other hand, ...-creases at first and then ....-creases.

2. Referring back to data in text: Estimate $\frac{dS}{dt}$ if there were 2 boys ill on the first day and 5 on the next day.
....

3. Estimate $a = -\frac{dS/dt}{SI}$ for the data in 2. if the total number of the boys in the boarding school remained 763.

4. If there were 1000 susceptible boys in the boarding school, instead of 763, would the threshold value still be 192? ....

Checking examples

- This whole section consists of the British boarding school example. Do not rush this example.

Problems

3, 4, 5, 6, 8

REVIEW OF CHAPTER 10

1. For exponential growth or decay the rate of change of y, that is $\dfrac{dy}{dx}$, is proportional to ....

2. What is the equilibrium solution for $\dfrac{d}{dt} = 0.2y$? .... Is it stable? ....

3. What is the equilibrium solution for $\dfrac{d}{dt} = -0.2(y - 30)$? .... Is it stable? ....

4. Newton's Law of heating and cooling says that:
   The rate of cooling = ... × ....
   ....

5. Translate into symbols and define all variables:

   • The rate at which a population grows is proportional to its size.
   ....
   ....

   • The rate at which people hear news is proportional to the number of people who have not heard the news.
   ....
   ....

   • The thickness of a layer of ice changes at a rate which is inversely proportional to the thickness of the ice.
   ....
   ....

Review Problems

2, 3, 5, 8, 16, 21, 27, 30

CHAPTER 11

GEOMETRIC SERIES

11.1 GEOMETRIC SERIES

CAREFULLY READ THROUGH THIS SECTION IN THE TEXTBOOK

Key points

- Introducing geometric series by means of practical examples.

- Finding the sum of a finite and an infinite geometric series.

Knowing the Book

1. The sum of a finite geometric series $a + ar + ar^2 + ar^3 + \ldots + ar^{n-1}$ is given by

   ....

2. The sum of an infinite series $a + ar + ar^2 + ar^3 + \ldots + ar^{n-1} + \ldots$ if $|r| < 1$ is given by ....

3. If $|r|$....... we say the series *diverges* and if $|r|$........ we say the series *converges*.

Checking the concepts

1. The geometric series with four terms and with $a = 10$ and $r = 0.5$ is given by:
   ....

2. For the geometric series $2 + 1 + 0.5 + 0.25 + \ldots$ we have $a =$............ and $r = $.......

3. The geometric series in 2. has a sum of ....

4. For geometric series $2 + 4 + 8 + 16 + \ldots$ we have $r = $ .... and the series is ....-vergent because ....

Checking examples

- **Examples 1 - 3:** These examples build on the preceding discussions on Drug Dosage and deposits into a savings account. First make sure that you follow the discussion.

- **Examples 4 & 5:** We normally deal with finite series; infinite series really only exist in theory although these examples both refer to "forever", meaning for the rest of your life. As is pointed out there is not much difference between a sum containing many terms and a sum of an infinite series.

Problems

2, 4, 14, 16, 20

11.2 APPLICATIONS TO BUSINESS AND ECONOMICS

CAREFULLY READ THROUGH THIS SECTION IN THE TEXTBOOK

Key points

- Looking at the following applications:

  - Present value of an annuity

  - The Multiplier effect

  - Market stabilization

Knowing the Book

1. An annuity is a sequence of .... payments made at .... intervals indefinitely or over a .... time.

2. The present value of annuity is the amount of money that must be deposited today to make a series of ....

3. The multiplier effect: Explain to yourself why the total effect of a rebate on the economy is much larger than the rebate itself.

4. Is the market stabilization point a *number of items* or a *point in time*?

Checking the concepts

1. An amount of $500 five years into the future, at an interest rate of 8%, compounded annually, has a present value of $................

2. If the interest was compounded continuously in 1. the present value would be $.............

3. An amount of $200 at present has a future value in four years, at a rate of 5% compounded annually, of $............

4. If the interest was compounded continuously in 3. the future value would be $................

Checking examples

- **Example 1:** This examples shows how to construct a geometric sequence to calculate the total value of an annuity.

- **Example 2:** An example to calculate the present value of an annuity. Note the 20th payment is made 19 years into the future because the first payment is made now.

- **Example 3:** An interesting example, showing again that "forever" often requires not much more than a fairly long period of time.

- **Example 4:** Say everyone spends 50% of what they earn and saves 50%. Does this have a dramatic effect on the total additional spending?

- **Example 5:** Another way of looking at it is that in the stable situation the 13 billion pennies produced per year must be equal to the 10% that is removed each year. Hence a total of 130 billion pennies in circulation.

Problems

1, 4, 8, 12

11.3   APPLICATIONS TO LIFE SCIENCES

CAREFULLY READ THROUGH THIS SECTION IN THE TEXTBOOK

Key points

- Looking at the following applications:

  - Repeated drug doses.

  - Accumulation of toxins in the body.

  - Depletion of natural resources.

- Geometric series versus differential equations.

Knowing the Book

1. At the steady state of a drug dosage, the quantity of the drug in the body varies between a .... level right after a dose is taken and a .... level right before a dose is taken.

2. We use a geometric series when a drug is given in .... doses and a differential equation when the drug is given in .... doses.

Checking the concepts

1. A person taking 100mg of a drug every hour of which 80% is eliminated during the hour has, right after tasking the second dose, ... mg in his blood.

2. The geometric series

$$5 + 5e^{-0.05} + 5e^{-0.10} + 5e^{-0.15} + \ldots$$

could represent the situation where .... mg of a drug is administered every day intravenously of which ....% is eliminated daily. The expression then gives ....

3. Geometric series or differential equation?

   (a) $100 is deposited once per year and grows at 5%, compounded yearly.

   (b) An income stream of $100 per year earns interest at 5%, compounded continuously.

4. Give in each of the cases in 3. the appropriate geometric series or differential equation.

   (a) ....
   (b) ....

Checking examples

- **Example 1:** Note that we work on units of 4 hours. Furthermore, the steady state can be calculated fairly easily by saying the at the beginning of a period the drug has to be supplemented by an additional 88% to get back to the steady state. So 200 mg is 88% and therefore 100% is 227.2727 mg.

- **Example 2:** The question here is what percentage of the substance remains after 12 hours. But first we need to find a formula for this exponentially decaying function. One option is to go for the formula $Q = Db^t$, another is to go for $Q = De^{kt}$. By using the half-life you can find the value of k and develop a geometric series. See Examples 2 & 5 where this approach was taken.

- **Example 3:** Note the term "continuous rate" which immediately alerts you to use the *e*-function. This points to a big difference between the economics examples and the life science examples. Annual payments are made at fixed intervals, whereas a drug leaves the body continuously. That is why we use the *e*-function here.

- **Example 4:** In this example we do not assume continuous consumption of oil, rather that it is used and accounted for at the end of the year. That is the reason why the factor to multiply by is 1.01 and not $e^{0.01}$.

- **Example 5:** The difference between "discrete" and "continuous", which are associated respectively with a geometric series and a differential equation, is illustrated here by either an injection every morning or an intravenous method. It would interesting be to sketch the graphs of the two cases.

Problems

1, 3, 6, 9, 11, 13

REVIEW OF CHAPTER 11

1. Give a geometric series for each of the following cases:

   - An amount of $100 increases at 5% per year for four years.

   - An amount of $100 increases at a continuous rate of 5% per year for four years.
   ....

2. The geometric series that gives the present value of an annuity of $1000 that is paid out every year for 10 years is: (Interest rate is 6% and the first payment is immediately).
   ....

3. Money is deposited monthly and earns 10% interest. Is the resulting geometric series convergent or divergent? ....

4. A government rebate of $5 billion is given and every individual saves 30% and spends 70%. The geometric series for the total additional spending is:
   ....

5. After 24 hours there is 30% left of a drug in the body before the new dose of 100 mg is given. The geometric series for the steady level of the drug in the body, right after taking a dose, is:

....

Review Problems

1, 5, 9, 11, 13

APPENDIX A

SELECTED ANSWERS

CHAPTER 1

1.1 WHAT IS A FUNCTION?
Checking the concepts
1. (a), (b) and (d)

2. .

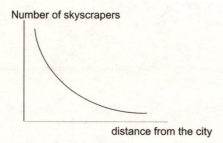

Number of skyscrapers

distance from the city

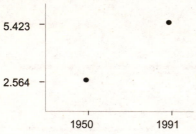

Population in millions

5.423 –

2.564 –

1950   1991

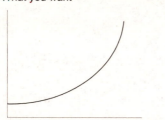

What you want

What you have

3. 1(b), 2(a), 3(d), 4(c).

4.
- $B = f(A)$
- $Q = f(H)$
- $P = f(S)$

Checking examples
- **Example 1:** Because we know that $f(500) = 100$ and so $c > 500$. The function values decrease as c increases.

1.2 LINEAR FUNCTIONS
Checking the concepts
1. 2000   5

2. $50^0 C$

3. difference

4. 1

5. $\frac{s}{r}$

6. decreasing

Checking examples
- **Example 1:** 220 sec

- **Example 2:** The waste grows at 1.875 million tons every year.
  The population increases and will produce more and more waste per year. The waste graph will probably not be a straight line.

1.3 RATES OF CHANGE
Checking the concepts
1. (a) change

   (b) rate of change

   (c) change

   (d) rate of change

2.
- $\frac{change\ in\ temperature}{change\ in\ time}$
- $\frac{change\ in\ price}{change\ in\ time}$

3. -10 sheep per year   de   -16 sheep per year   -13 sheep per year

4. de

5. in

6. down

Checking examples
* **Example 3:** parts per million   mm
de-   0.28mm   0.22mm
-0.06 mm   152 ppm
$\frac{-0.06}{152}$ mm/ppm = -0.000395 mm/ppm
de-   0.000395 mm

1.4 APPLICATIONS OF FUNCTIONS TO ECONOMICS
Checking the concepts
1. (a), (b) and (c)

2. $C(q) = 200 + 7q$

3. $R(q) = 7q$

4. (b) and (d)

5. initial cost   depreciation per year
$7500

6. increasing
increasing
decreasing
increasing
decreasing

7. The additional cost of producing one more item than 100 is $2000.

Checking examples
* 2. The equilibrium price is $36.

* 3. The point (2, 10).

1.5 EXPONENTIAL FUNCTIONS
Checking the concepts
1. 2.5   2

2. 25
25   30
30   36

3. y increases by a factor of 1.5, so the missing values are 2.7, 4.05 and 6.075.
y increases by a constant of 0.6, so the missing values are 2.4, 3.0 and 3.6.

4. $P = 100(1.12)^t$

5. (a) F   3.5% = 0.035
(b) F   It is decreasing.
(c) F   It is called a power function.
(d) T
(e) F   Both are concave up.

6. 1. (c)   2. (d)   3. (a)   4. (b)

7. 7.389   1.030   1.649   0.698

1.6 THE NATURAL LOGARITHM
Checking the concepts
1.
* 2.1
* 0.8
* 2.8
* 1.2
* 3.7
(These are all estimates)

* 2.1
* 0.8
* 2.8
* 1.2
* 3.7

2. 3.912   50

3. 4.263   1.45

4. (a) F ln e =1
(b) F ln 200 = ln 100 + ln 2
(c) T
(d) T
(e) F ln$\frac{10}{3}$ = ln 10 - ln 3
(f) T
(g) T
(h) F ln$(5e^{3t})$ = ln 5 + ln e^{3t} = ln 5 + 3t
(i) T
(j) F $e^3 e^4 = e^7$

5. (a) ln 1.04   0.03922
(b) 1.0565

6. a - (ii)   b - (iv)   c- (iii)   d - (i)

7. 30   5

8. in   up   in   up

1.7 EXPONENTIAL GROWTH AND DECAY
Checking the concepts

1. 500   20
   250   20

2. Half-life   Doubling time

3. T   T   T

4. 0.6931   (ln 2)

5. 2000   8   annually
   2000   8   continuously

6. T   T

7. The doubling time of an exponential function e^{rt} is given by $\frac{\ln 2}{r}$. But $\ln 2 \simeq 0.7$ and if the number value of the growth percentage is taken (eg. 7 instead of 0.07) then we multiply $\ln 2$ by 100.

1.8 NEW FUNCTIONS FROM OLD
Checking the concepts

- 1. F   There are exceptions such as a constant function.

2. T

3. T

4. F $f(g(t)) = t$

5. F   It is first reflected and then shifted up 3 units.

6. F   True if $f = g$ or if $f = g^{-1}$.

1.9 PROPORTIONALITY, POWER FUNCTIONS AND POLYNOMIALS
Checking the concepts

1. (a)   (c)   (c)

2. (a)

3. The first two.

4. • $10^{\frac{1}{3}}$
    • x^2

5. $y = \frac{k}{\sqrt{t}}$

6. (a) F   Polynomial powers can only be positive.
   (b) F   Polynomial powers can only be integral.
   (c) T
   (d) F   Four times.
   (e) T
   (f) T
   (g) T

Checking examples

- **Example 1:** Straight line through the origin.

- **Example 5:** 48   96   See Figure 2.

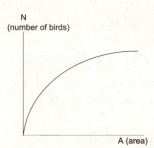

1.10 PERIODIC FUNCTIONS
Checking the concepts

1. 3.5   9

2. $2.3 + 2\pi = 8.58$

3. $\frac{2\pi}{4} = \frac{\pi}{2}$
   $\frac{2\pi}{\frac{1}{4}} = 8\pi$

4. -0.866   0.866

5. 12   120   -80

6. $y = 5\sin\frac{\pi}{3}t + 25$

REVIEW OF CHAPTER 1

1. (a) An exponential function
    $P = 4.3e^{0.0157t}$

    (b) A sine function (periodic)
    $V = 12 + 2\sin\frac{\pi}{6}t$, t in months

    (c) An exponential function
    $P(t) = 350(1.055)^t$

    (d) A cubic polynomial
    $y = ax^3 + bx^2 + cx + d$

    (e) A linear function
    $y = 100 - \frac{40}{3}t$, t in hours.

2. $\frac{P(t_1) - P(t_0)}{t_1 - t_0}$

3. $y = x^3$

4. 80 tons

5. 100 items can be produced at a price of $2.50

6. 32   40.4
    53.1   79.65

7. equilibrium point   zero

8. initial population   in   increases by
    1   3.5%

FOCUS ON MODELING

FITTING FORMULAS TO DATA

Checking the concepts

1. (c)

2. 7   7   pound   price   7

3. $30.50 $79.50

4. (d)

5. 100%

Checking examples

- **Example 1:**

    - $145250

    - $16500 $133500

- **Example 2:**

- $707.86

- Approximately $3.612

COMPOUND INTEREST AND THE NUMBER e

Checking the concepts

1. 6.2315%

2. $105.0625 $105.0945 $105.1162
   100   5%   six-monthly
   $105.0625   1
   100   5%   quarterly $105.0945
   1
   5.127

3. 1.08328
   8.328%

Checking examples

- **Example 4:** Yes (7.19%)

- **Example 6:** 1.072508

FOCUS ON THEORY

LIMITS TO INFINITY AND END BEHAVIOR

Checking the concepts

1. (a)   (a)   (a)   (a)   (a)

2. ∞ $-\infty$ 0

3. x^9

CHAPTER 2

2.1 INSTANTANEOUS RATE OF CHANGE

Checking the concepts

1. (a), (c) and (e)

2. $1.49 \leq x \leq 1.51$
   $1.499 \leq x \leq 1.501$
   $1.4999 \leq x \leq 1.5001$

3. 0.02   0.1   0.002

4. $0 \leq x \leq 0.01$ (for example)
   $-0.0001 \leq x \leq 0$
   $1.9999 \leq x \leq 2$
   $2 \leq x \leq 2.0001$
   $1.3999 \leq x \leq 1.4$

5. (a) -0.4396
   (b) -2.75013
       The function is decreasing
       (exponential decay.)

6. 0.353664
   0.353554
   0.353553
   0.35355

7. 1 (d)   2 (c)   3 (a)   4 (b)

8. tangent   connecting

9. nega   -3 or -5.6 or -4.3

10. (a) $P'(4) = 2.34$
    (b) $f'(5) = 0$
    (c) $S'(3) = 5$

Checking examples

- **Example 1:** -1.934   -1.730   -1.934
  -1.730   -1.832
  -1.849   -1.808   -1.849   -1.808
  -1.828

- **Example 6:**

  1. 0.0157   0.157

2.2 THE DERIVATIVE FUNCTION

Checking the concepts

1.
3	the average rate of change of f over $0 \leq x \leq 3$
$2x$	the derivative function
6	the instantaneous rate of change of f at 3
4	the derivative of f at 2

2.
A	positive	0.6	above
B	negative	-7	below
C	zero	0	cuts the x-axis
D	positive	1.5	above
E	zero	0	cuts the x-axis
F	negative	-1.5	below

3. (a) 3.   (b) 2.   (c) 4.   (d) 1.

Checking examples

- **Example 1 & 2:** The derivative of the derivative is a line with positive slope that cuts the x-axis at 2.

2.3 INTERPRETATIONS OF THE DERIVATIVE

Checking the concepts

1. if the x value increases by 1 unit (from 5 to 6) then the y value *increases* by approximately 1.6 units.
   if the x value increases by 1 unit (from 10 to 11) then the y value *decreases* by approximately 2.5 units.

2. $2 + 2 \times 1.6 = 5.2$
   $5 + 2 \times (-2.5) = 0$

Checking examples

- **Example 2:**

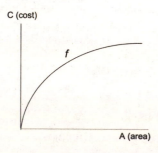

1000   60000   50   70000

- **Example 3:** that it will cost approximately $90 more to extract 1 ton more when 2500 tons have already been extracted.

- **Example 4:** f is concave down because f is increasing at a decreasing rate.

- **Example 5:** If 5mg is administered it will stay in the person's system for 2 hours. If another 1mg is administered (6mg) it will stay approximately 1 hour longer (3 hours) in the person's system.
  $f(8) = 3.5$
  $f'(8) = 1.2$

2.4 THE SECOND DERIVATIVE

Checking the concepts
1. .

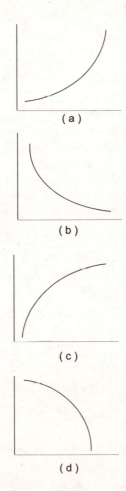

(a)

(b)

(c)

(d)

up   increasing
down   decreasing

2. (a) and (b)
   (c) and (d)

3. The logistic curve (p114) for example.

4. $f'(4) = 1.525$ in   up $>$

5. An estimate for $f''(2)$ is 0.09375.

Checking examples
- **Example 2:** the rate of change is a minimum at t^*.

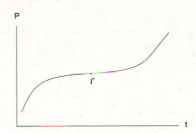

- **Example 3:** 2.

2.5 MARGINAL COST AND REVENUE

Checking the concepts
1. (a) $500
   (b) $10
       120      $18 Yes $8
   (c) $15 $10   de   will not
   (d) the cost of manufacturing one more item is the same as the revenue gained.

2. more
   less
   no more and no less

3. constant (the slope of a straight line is constant)

Checking examples

- **Example 1:** A loss is made.

- **Example 5:** negative   positive
  small   maximum   decreases

REVIEW OF CHAPTER 2

1. $1.249 \leq t \leq 1.251$
   $1.2499 \leq t \leq 1.2501$

2. limit of average   smaller and smaller
   intervals around a

3. 2.4022   2.4000   2.40

4. - the tangent at $x = 1$
   - the line connecting the points
     corresponding to $x = 1$ and $x = 3$

5. increasing   has a horizontal tangent

6. s increases from 12 to 13 N will
   increase by approximately 2.3

7. down

8. $y = \sin x$ between $x = 0$ and $x = 3\pi$

9. quantity   the 11th pair of shoes will
   cost approximately $25 to manufacture.

FOCUS ON THEORY
LIMITS, CONTINUITY AND THE DEFINITION OF THE DERIVATIVE

Checking the concepts

1. $h + 4$ 4

2. The term $\frac{2+h}{h}$ does not approach a fixed
   number as h approaches zero.

3. No, the function has a break at $x = 1$

4. 57   34

CHAPTER 3

3.1 DERIVATIVE FORMULAS FOR POWERS AND POLYNOMIALS
Checking the concepts

1. 6   approx 2   approx 6

2. 6   decrease   6   increase   6

3. 0.5

4. (a) Wrong. $\frac{d}{dx}(2)^5 = 0$.
   (b) Correct.
   (c) Wrong.   No rule in this section
       applies.
   (d) Wrong. $\frac{d}{dx}x^{-\frac{1}{2}} = \frac{-1}{2}x^{-\frac{3}{2}}$

5. Yes.

6. difference   derivatives
   constant   the derivative of the function

Checking examples

- **Example 8:** The second derivative is the
  derivative of the first derivative. The
  second derivative is negative to the left of
  B and positive to the right of B. The
  graph of the second derivative is a line
  that cuts the x-axis at B from below.

3.2 EXPONENTIAL AND LOGARITHMIC FUNCTIONS
Checking the concepts

1. (a) $f(x) = e^x$
   (b) $f(x) = \ln x$

2. below   below   above

3. The function $= (\frac{1}{2})^x$ is a decreasing
   function and therefore the slope is always
   negative.

4. (a) Wrong. The derivative of a constant
       is zero.
   (b) Wrong. The answer is $(2.3)^x \ln(2.3)$

(c) Wrong. The derivative of a constant is zero.

5. in   de

6. in   above   de   in   below

Checking examples
- **Example 3:** 3.73598 million people per year   10 235 people per day

- **Example 4:** $y = 100x - 5.605$

3.3 THE CHAIN RULE
Checking the concepts
1. (a) $5(z)^4 \frac{dz}{dt}$

    (b) $\frac{1}{2}(f(t))^{-\frac{1}{2}} f'(t)$

    (c) $f'(z(t)) z'(t)$

    (d) $z(t)$

    (e) $z(t)$ $\frac{dv}{dz}$

2. (a) Wrong $-0.01 e^{-0.01x}$

    (b) Wrong $4(x^3 + x)^3 (3x^2 + 1)$

    (c) Wrong $(1 + \frac{x}{10})^9$

    (d) Wrong $-4(1 - t)^3$

3. $e^{z(x)} z'(x)$

Checking examples
- **Example 6:** (a) $f(t) = 1000(1.08)^t$
  (b) $f(10) = 2158.92$
  $f'(t) = 1000(1.08)^t \ln 1.08$ and
  $f'(10) = 166.15$

3.4 THE PRODUCT AND QUOTIENT RULES
Checking the concepts
1. Yes $\frac{d}{dx}(5) \times x^2 + 5 \times \frac{d}{dx}(x^2) = 0 + 10x^2$

2. (a) $v'(t)w(t) + v(t)w'(t)$

    (b) $2s(t)s'(t)$

    (c) $f(x) + xf'(x)$

    (d) $2pf(p) + p^2 f'(p)$

3. (a) $\frac{v'(t)w(t) - v(t)w'(t)}{w^2(t)}$

    (b) $\frac{0 \cdot s(t) - s'(t)}{s^2(t)} = \frac{-s'(t)}{s^2(t)}$

    (c) $\frac{f(x) - xf'(x)}{f^2(x)}$

Checking examples
- **Example 3:** $R'(21) = 70.38$ which means that if one more item is sold, the revenue will increase by approximately 70.38. So more items should be sold, even at a lower price.
  $R'(693) = -10.79$. If one more item is sold, revenue decreaes by about 10.79. It is better to sell less at a higher price.

3.5 DERIVATIVES OF PERIODIC FUNCTIONS

1. 0.707   -0.707

2. 3.943

3. cuts the x-axis

4. in

REVIEW OF CHAPTER 3
1.
   - nx^{n-1}
   - $-nx^{-n-1}$
   - $\frac{1}{2}(x^n)^{-\frac{1}{2}} nx^{n-1}$
   - $\frac{1}{n} x^{\frac{1}{n}-1}$

2.
   - $a > 1$
   - $a < 1$

3. (a) pos   de
   (b) pos   in

4. (a) $-2(u(t))^{-3} \frac{du}{dt}$

   (b) $\cos u(t) \frac{du}{dt}$

   (c) $6(u(t))^5 \frac{du}{dt}$

   (d) $\frac{u'(t)}{u(t)}$

5. $R = 20e^{-0.02q} q$
   $20e^{-0.02q} - 0.4qe^{-0.02q}$

6. Chain rule: $\frac{d}{dx}(x^{-1}) = -x^{-2}$
   Quotient rule: $\frac{d}{dx}(\frac{1}{x}) = \frac{0-1}{x^2} = \frac{-1}{x^2}$

FOCUS ON THEORY

VERIFYING THE DERIVATIVE FORMULAS

Checking the concepts

1. (a) 3

   (b) $e^{3(x+h)}$

   (c) $(x+h)^2+(x+h)=x^2+2hx+h^2+x+h$

   (d) $\frac{1}{x+h}$

2. derivative definition

3. (a) $\frac{1}{x}$

   (b) 0

   (c) x^2

   (d) x

   (e) 1

   (f) 3

4. $\frac{e^{3(x+h)}-e^{3x}}{h}$ $e^{3x}\left(\frac{e^{3h}-1}{h}\right)$ $3e^{3x}$

Checking examples

- **Example 3:** 1.005017   1.00005

CHAPTER 4

4.1 LOCAL MAXIMA AND MINIMA

Checking the concepts

1. - T $f'(2)=0$
   - F 0 is not in the domain of f, so $x=0$ cannot be a critical point.
   - T
   - T
   - T

2. minimum   maximum

3. Yes, at $x=1$.

4. $y=x^3$ is a good example

5. - F   Look at $y=x^3$ at $x=0$.
   - T
   - F   Again look at $y=x^3$ at $x=0$.

Checking examples

- **Example 3:** $y=x^3$, once again.

4.2 INFLECTION POINTS

Checking the concepts

1. The sine function.

2.
0	0
1	0
4	1
0	1
0	0
3	2

3. - T
   - F Look at $y=x^4$ at $x=0$.

Checking examples

Example 5: The graph would repeat itself upwards all the way. It has three inflection points.

4.3 GLOBAL MAXIMA AND MINIMA

Checking the concepts

1.
   - F Example: $\frac{1}{x}$ on (0,1]
   - T
   - T
   - T
   - F f has neither a global maximum nor a global minimum.
   - T

2. (a) → 1.
   (b) → 3.
   (d) → 2.

3. 6   3   2   4

4. Yes No

5. No, these can occur at the endpoints.

6. Yes   Yes

4.4 PROFIT, COST AND REVENUE

Checking the concepts

1. parallel   intersect

2. should (The revenue will increase by approx $2.5 whereas cost will increase by approx $2.0 for the next unit produced.)

3. $q(80 - 3q)$ $p(-\frac{1}{3}(p - 80))$
   both function have a maximum

Checking examples
   - **Example 1:** de   in   de   in

4.5 AVERAGE COST

Checking the concepts

1. dollars per item
   dollars per item

2. increases

3. (a) 6
   (b) Yes

4.
   - $52400

   - $192000
   Yes

Checking examples
   - **Example 2:** The average cost will decrease.

4.6 ELASTICITY OF DEMAND

Checking the concepts

1. No

2. No units; it is a ratio.

3. If the price increases by 1% the demand will decrease by 2%

4.
   - decrease by 4% to 48.
   - increase by 4% to 52.
   - decrease by 0.5% to 49.75.
   - increase by 0.5% to 50.25.

5.
   - $1000
   - de $969.6   lowered
   - in $1004.95   raised

4.7 LOGISTIC GROWTH

Checking the concepts

1. (a) → 3
   (b) → 1
   (c) → 2
   (d) → 5
   (e) → 4

2. 640

3. No, no population can grow from nothing. Also from the formula, if $t = 0$ then $P \neq 0$.

4. First mark the percentages on the $y - axis$ and then move over to the $x - axis$

5. 6mg   8 or 9mg

6. (a) 5mg   (b) 8mg

Checking examples
   - **Example 1:** $\frac{4}{101}$

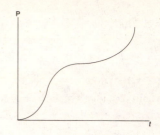

- **Example 2:** See sketch at top of page.

- **Example 4:** Probably not. People will always buy new CD players - old ones break and new people are born.

4.8 THE SURGE FUNCTION AND DRUG CONCENTRATION

Checking the concepts

1. • 0.25   3
   • $t = 4$
   • 4.4146

2. hours   mg/ml

3. (b)   (a)

4. (b)

5. (a) after 0.2 hours
   (b) after 4 hours
   (c) 3.8 hours

Checking examples

- **Example 2:** Harm (It would slow down the absorption even more.)

- **Example 4:** The drug will have to be administered more frequently.

REVIEW OF CHAPTER 4

1. (a) F The critical point could be an inflection point.
   (b) T
   (c) T
   (d) F It should be decreased.

2. local maximum or a local minimum

3. Parabolic shape with a minimum point

4. Compare function values at the local maxima and at the endpoints of the interval.

5. up   down

6. in   de

7. • A surge function
     $E = ate^{-t}$
   • A logistic curve
     $I = \frac{400000}{1 + Me^{-kt}}$

CHAPTER 5

5.1 ACCUMULATED CHANGE

Checking the concepts

1. 5 ft   9 ft

2. 5.75 ft $(5 \times 0.5 + 6.5 \times 0.5)$
   7.75 ft $(6.5 \times 0.5 + 9 \times 0.5)$

3. .

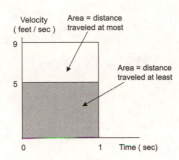

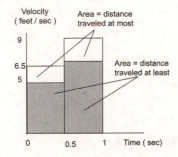

The function lies below the x-axis between $x = 0$ and $x = 2$, so the integral is negative.

4. 44 ft   76 ft

5. (a)   4.5 inches
   between 4 inches and 6 inches

5.2 THE DEFINITE INTEGRAL

Checking the concepts

1. in   improve

2. 4   2   2.5   3   3.5   4

3. 0.2   3   3.2   3.4   3.6   3.8 ... 4.8   5

4. 7   9.25   12   15.25   19

5. .

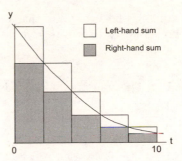

Left-hand sum   upper estimate
Right-hand sum   lower estimate

6. (a) T

   (b) T

   (c) T

   (d) F (only for decreasing or increasing functions)

   (e) T

5.3 THE DEFINITE INTEGRAL AS AREA

Checking the concepts

1. above $\int_0^2 1.3^t dt$

2. (a) .

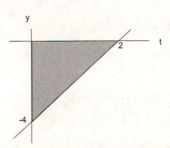

The function lies below the x-axis, so the integral on this interval is negative.

(b) The first expression represents an area above the x-axis and the second expression represents the sum of areas above and below the x-axis.

(c) 8

3. 0   3
   3   6
   take the absolute values of the two
   integrals and add them together.

4. • T

    • F   The sum of two integrals can be
    zero.

    • T

    • T

    • F   An area is always positive.

5.4 INTERPRETATIONS OF THE DEFINITE INTEGRAL

Checking the concepts

1. inches
   feet per sec
   dollars
   items

2. 1d   2a   3b   4c

3. • T

    • T

    • F   The area under the graph is not
    zero.

Checking examples

• **Example 3:** $t = 1$ $t = 3.75$ $t = 2$.

5.5 THE FUNDAMENTAL THEOREM OF CALCULUS

Checking the concepts

1. (a) $F(4) - F(2)$

    (b) $2(5)^2 - 2(3)^2 = 32$

    (c) $F(2)$ $F'(t)$

    (d) $5(3)^2 - 5(0)^2 = 45$

    (e) $F(7)$

2. in

3. The approximate cost of producing the
   41st item is 5 (dollars).
   The cost of increasing production from 40
   to 50 items is 55.
   The variable cost of producing the first 50
   items is 105.
   The total cost of producing the first 50
   items is 150.

REVIEW OF CHAPTER 5

1. 4   3   3.25   3.5   3.75   4

2. $0.25(f(3) + f(3.25) + f(3.5) + f(3.75))$

3. • A right-hand sum

    • Area included between the function
      and the x-axis

    • A left-hand sum

4. total change

5. The area above the x-axis is the same as
   the area below the x-axis.

6. $\int_0^1 (x - x^3)dx + \int_1^2 (x^3 - x)dx$

7. the cost of increasing production from 100
   to 250 items
   dollars

8. • Riemann-sum (left or right or
      average)

    • Riemann-sum
      Technology
      Fundamental Theorem (if possible)

FOCUS ON THEORY

THEOREMS ABOUT DEFINITE INTEGRALS

Checking the concepts

1. Because f is the derivative of G, the
   change in G over the interval $[x, x + h]$ is
   given by the area under f over this
   interval. This area is approximated by
   $f(x)h$.

2. 0   0.5   2   4.5
   Yes, the slope is positive and increasing.

3.
- F   No such rule.
- T
- T
- F   No such a rule
- T

CHAPTER 6

6.1 AVERAGE VALUES
Checking the concepts
1. (1) C   (2) A   (3) B

2. A fair estimate for the height is 0.3

3. Yes, for a negative function, for example.

4.
- T
- F (an integral is calculated, not an area but a difference of areas)
- F (it is less than $\frac{1}{2}$)

5.
- 2
- 0   20

6.2 CONSUMER AND PRODUCER SURPLUS
Checking the concepts
1. At a price of $50 consumers will be willing to buy 200 items.
   At a price of $20 producers can supply 200 items.
   less
   more

2. $1.83 (1.832)
   $\int_0^{1000} 100e^{-0.004t} dt - 1832$

3. This is the amount that the producers gained by selling at a higher price than what they were willing to accept.

4. This is what the consumers gained by buying at a lower price than what they were willing to buy at.

6.3 PRESENT AND FUTURE VALUES
Checking the concepts
1. $600 $100

2. $67492.94

3. $9417.65

4. 0   1200

5. 110   120

$$\int_0^1 (100 + 10t)e^{-0.06t}\,dt$$

$$\int_1^2 (100 + 10t)e^{-0.06t}\,dt$$

6. $e^{0.12} \int_0^2 (100 + 10t)e^{-0.06t}\,dt$

6.4 RELATIVE GROWTH RATES

Checking the concepts

1.
   - $\frac{dP}{dt} = 10e^{0.1t}$
   - 0.1 or 10%

2.
   - $\int 0.1\,dt = 0.3$ 0.3
     $e^{0.3} = 1.3499$ 34.99

3. $\ln P(5)$ $\ln P(0)$

4. 200%

5. $\int_0^{50} \frac{P'(t)}{P(t)}\,dt$

6. $e^{0.3}$

REVIEW OF CHAPTER 6

1. 150

2.
   - between the demand curve and the line at $p* = 10$.
   - The consumers gain $1000 by buying at $10 than by buying at what they were willing to pay.
   - between the supply curve and the line at $p* = 10$
   - The producers gain $1500 by selling at $10 than by selling at what they were willing to sell at.

3. $1000e^{-0.4} = \$670.32$

4. $\int_0^{10} 1000e^{0.06t}\,dt = 13701.98$

5.
   - The population increased by 20 over the three year period.
   - The factor is $e^{0.2}$ so the percentage increase is 22.14%.

CHAPTER 7

7.1 CONSTRUCTING ANTIDERIVATIVES ANALYTICALLY

Checking the concepts

1. (a) T
   (b) T
   (c) F an antiderivative of $8x^3$ is $2x^4$.
   (d) F An antiderivative of e^{2x} is $\frac{1}{2}e^{2x}$.
   (e) T

2. (a)) F If you differentiate the right-hand side, you should use the product rule.
   (b) T
   (c) F $\int \frac{1}{x}\,dx = \ln x + C$
   (d) T

7.2 INTEGRATION BY SUBSTITUTION

Checking the concepts

1. $w = x^2 + 1$
   $w = x^3$
   $w = x^2 + 1$
   $w = 3 - t$

2. (b)   (d)   (e)

3. $2x$
   -1
   $-2\sin 2x$

7.3 USING THE FUNDAMENTAL THEOREM TO FIND DEFINITE INTEGRALS

Checking the concepts

1. $1000e^{0.01t}$ $1000e^{0.01} - 1000$
   $-\frac{3}{2}\cos 2t$ 0
   $\frac{1}{4}x^4$ 16
   $2\ln x$ $2\ln 2$
   $\frac{-1}{2x^2}$ $\frac{3}{8}$

2. 10

7.4 ANALYZING ANTIDERIVATIVES GRAPHICALLY AND NUMERICALLY

Checking the concepts

1. 1 (C)   2 (D)   3 (B)   4 (E)
   5 (A)

2. in   in
   in   de

3. • 7
   • 4

REVIEW OF CHAPTER 7

1. • $\frac{3}{2}x^2 + x + C$
   • $\frac{-4}{3}\cos 3t + C$
   • $e^{t+4} + C$

2. • $w = \sqrt{x}$
   • $w = 4 + \cos\theta$
   • $w = x^2 + 4$

3. 4   0 $f(x) = 10e^{0.1t}$

4. The antiderivative

   • increases at a decreasing rate
   • is linear
   • changes from decreasing to increasing

CHAPTER 8

8.1 DENSITY FUNCTIONS

Checking the concepts

1. The percentage of the population between the ages of 20 and 30.

2. height in inches
   fraction of adult females per inch
   the percentage of females with height between the x-values.

3. The fraction of adult females with height between 60 and 62 inches.

Checking examples

• **Example 2:** Approximately
   $5\frac{2}{3} \times 0.5 = 0.2833$ or 28.33%.
   Approximately 1×0.05 or 5%.

8.2 CUMULATIVE DISTRIBUTION FUNCTIONS AND PROBABILITY

Checking the concepts

1. The percentage of people younger than 6 years.

2. $P(5) - P(3)$

3. • $p(t)$ is a horizontal line with height 0.1 and $P(t)$ is a line through the origin with slope 0.1 until it meets (10,1) and then it is a horizontal line $P(t) = 1$ for $t \geq 10$,

   • $0.1t, 0 \leq t \leq 10$

   • 0

   • 1   1

   • $P(4) - P(2) = 0.4 - 0.2 = 0.2$

   • The fraction of numbers t that will be less than 6

4. $1 - \int_0^{60} h(t)dt$

8.3 THE MEDIAN AND THE MEAN
Checking the concepts
1. 5   4.4

2. horizontal

3. half

4. balance

Checking examples
- **Example 1:** If the function $p(t)$ is extended it becomes negative. The integral up to 85.35 will consist of a negative part and a positive part and that will add up to 0.5 but of course does not apply.

REVIEW OF CHAPTER 8
1.
   - time in minutes   fraction of population per unit time
   - the fraction of population that will wait between 3 and 5 minutes.
   - the fraction of population that will wait for less than 5 minutes.
   - The median is 3.

2. 1(d)   2(a)   3(c)   4(b)

CHAPTER 9

9.1 UNDERSTANDING FUNCTIONS OF TWO VARIABLES
Checking the concepts
1.
   - $H = f(a, s)$ where H is the height of the oak tree, a is the age and s is the fertility of the soil.
   - $C = f(t, i)$ where C is your ability to concentrate, t is the time of day and i indicates how interesting the topic is.

2.
   - $C^2 + Ct$, for example.
   - $rg - r^3 + 3g$, for example.

3. $P = f(L, k, t)$

4.
   - the size of a population of initially 2.5 million, 10 years later.
   - what the sizes of populations with different initial sizes will be 15 years later.
   - how the size of a population of initially 4 million will develop with time.
   - $P = Ie^{0.02t}$.

9.2 CONTOUR DIAGRAMS
Checking the concepts
1. At a vertical waterfall two contours can touch. This would, however, not be a function as there can be only one function value at every point in the domain.

2. Temperatures change gradually.

3. Lines through the origin but not including the origin.

9.3 PARTIAL DERIVATIVES
Checking the concepts
1. $f_s(s, g)$ $f_g(s, g)$

2. You will gain (approximately) $2.50 by investing one extra dollar if you have invested $1000 for two years.
You will gain (approximately) $30.50 by waiting one more year if you have invested $1000 for two years.

3.
   - 28.5
   - 37
   - 30.5

Checking examples

   - **Example 4:** $-0.4°F/feet$. It means that if you move 1 foot away the temperature will decreases by $-0.4°F$
   $0.8°F/minute$. It means that if you wait one more minute the temperature will increase by approx $0.8°F$

9.4 COMPUTING PARTIAL DERIVATIVES ALGEBRAICALLY

Checking the concepts

1. (a) F   It could be a function of y.
   (b) T
   (c) T
   (d) F   The powers must be fractional.

2. If you increase the current 50 workers by 1, production will increase by approx 2. If you increase the current value of equipment of 100 million by 1 million, production will increase by approx 20.

Checking examples

   - **Example 4:** After 2 hours the concentration decreases by -0.5413 units per unit increase in initial dose.
   After 1 hour and a dosage of 1 mg, the concentration could be at a maximum.

9.5 CRITICAL POINTS AND OPTIMIZATION

Checking the concepts

1. No, $f_y(0,1) \neq 0$

2. Yes, $f_s(1,2) = f_t(1,2) = 0$

3. Yes, the global minimum can be on the boundary.

4. (a) F It can also occur at a boundary
   (b) F It can occur on the boundary.
   (c) F $f(x,y) = x^3y^3$ has no global maximum.

5. It is clear that the function values can get no smaller than 0.

Checking examples

   - **Example 2:** It is clear that the values on the boundary are smaller than the function value at the local maxima.

9.6 CONSTRAINED OPTIMIZATION

Checking the concepts

1. $20x + 15y \leq 500$

2.
   - .

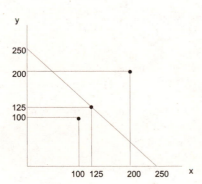

   - $2000   is   below
   - $4000   isn't   above
   - (125, 125)   on

3. a maximum of 1300 units can be produced and for that you have to buy 12 units of x and 13 units of y.

4. $1   0.034 units

REVIEW OF CHAPTER 9

1. • $F = f(r, s)$ where F is the amount of food grown, r the amount of rain and s the amount of fertilizer used (for example).

    • $C = f(t, p)$ where ....

    • $M = f(A, r, t)$ where ....

2. (a) a park of 8000 square miles and rainfall of 19 inches per year can carry 7000 elephants.

    (b) if the area of the park is increased from 8000 to 9000 square miles it can carry another 1000 elephants (approx).

    (c) if the rainfall increases from 19 to 20 inches per year the park can carry another 100 elephants (approx).

3. (a) positive
   (b) decreasing

4. de   de

5. $P = kN^\alpha V^\beta$

6. $f_x(3, 4) = 0$ and $f_y(3, 4) = 0$ either of $f_x(3, 4)$ or $f_y(3, 4)$ does not exist.

7. either using the Second Derivative Test or by checking the function values surrounding it

8. (a) quantities

    (b) the number of items produced

    (c) See "Method of Lagrange Multipliers," p359 in textbook.

    (d) if the budget increases by \$1 then production increases by 0.004 units (approx)

CHAPTER 10

10.1 MATHEMATICAL MODELING: SETTING UP A DIFFERENTIAL EQUATION?

Checking the concepts

1. • Decays at a rate of 1.6 million per year

    • We do not know the function P in terms of t. This is what we need to solve.

    • 12.5 million

2. • $\dfrac{dP}{dt} = kP$

    • $\dfrac{dP}{dt} = 0.05P$

    • $\dfrac{dP}{dt} = k(10000 - P)$

    • $\dfrac{dP}{dt} = -0.02P$

10.2 SOLUTIONS OF DIFFERENTIAL EQUATIONS

Checking the concepts

1. 14

2. $P(t) = 100e^{0.1t}$

3. (b)

Checking examples
   • **Example 2:** Yes

10.3 SLOPE FIELDS

Checking the concepts

1. -1

2. $x - y = x - (x - 1) = 1$. The slope is 1 on this line.

3. x
   y

4. 1C, 2B, 3A

10.4 EXPONENTIAL GROWTH AND DECAY

Checking the concepts

1. in-   10

2. in-   200

3. $Q = Q_0 e^{-0.05t}$ $Q = 250 e^{-0.05t}$

4. $\dfrac{dy}{dt} = -0.1y$

   $\dfrac{dy}{dt} = 0.025y$

5. concentration

   quantity   time

6. $\dfrac{d}{dt}Q = -kQ$

10.5 APPLICATIONS AND MODELING

Checking the concepts

1. $y = 10 + Ce^{2t}$ $y = 10 - 4e^{2t}$

2. $y = 10$

3. The cup with the higher temperature.

4.
   - $70°F$
   - 13 degrees per minute
   - 8 degrees per minute
   - $y = 70 + Ce^{-0.1t}$
   - $y = 70 + 130e^{-0.1t}$
   - $117.8°F$
   - $y = 70$. It is stable.

5. It goes to zero.

10.6 MODELING THE INTERACTION OF TWO POPULATIONS

Checking the concepts

1. $\dfrac{dr}{dt} = -0.02r$

   $\dfrac{dw}{dt} = 0.5w$

   The term krw is added to the first equation and the term crw is subtracted from the second equation (where k and c are two arbitrary positive constants).

2. The robins will eventually cause the worm population to decrease.

3. 1   2

4. A trajectory gives the relationship between the two population numbers whereas the solution present two functions of time.

10.7 MODELING THE SPREAD OF A DISEASE

Checking the concepts

1. de, in, in, de

2. $\dfrac{2-5}{1} = -3$

3. 0.00197 (using $s = 761$ and $I = 2$)

4. Yes, the threshold is independent of the initial number of susceptible boys.

REVIEW OF CHAPTER 10

1. y

2. $y = 0$ No

3. $y = 30$ Yes

4. a constant   difference in temperature between object and surroundings.

5.
   - $\dfrac{dP}{dt} = kP$ where P is the size of the population.
   - $\dfrac{dN}{dt} = k(L - N)$ where L is the total number of people in the population and N the number of people who have heard the news.
   - $\dfrac{dT}{dt} = \dfrac{k}{T}$ where T is the thickness of the layer of ice.

CHAPTER 11

11.1 GEOMETRIC SERIES
Checking the concepts
1. $10 + 5 + 2.5 + 1.25$

2. 2   0.5

3. 4

4. 2   di $r \geq 1$

11.2 APPLICATIONS TO BUSINESS AND ECONOMICS
Checking the concepts
1. $500(1.08)^{-5}$

2. $500e^{-0.40}$

3. $200(1.05)^4$

4. $200e^{0.20}$

11.3 APPLICATIONS TO LIFE SCIENCES
Checking the concepts
1. $100 + 100 \times 0.2$

2. 5   5   total amount of the drug in the body in the long run.

3. (a) geometric (b) differential equation

4. (a) $100 + 100(1.05) + 100(1.05)^2 + \ldots$
   (b) $\frac{dP}{dt} = 0.05P + 100$

REVIEW OF CHAPTER 11
1.
   - $100 + 100(1.05) + 100(1.05)^2 + 100(1.05)^3$
   - $100 + 100e^{0.05} + 100e^{0.10} + 100e^{0.15}$

2. $1000 + 1000e^{-0.06} + 1000e^{-0.12} + \ldots 1000e^{-0.54}$

3. Divergent

4. $5 + 5 \times 0.70 + 5 \times (0.70)^2 + 5 \times (0.70)^3 + \ldots$

5. $100 + 100(0.30) + 100(0.30)^2 + 100(0.30)^3 + \ldots$